ALSO BY BILL GATES

The Road Ahead (with Nathan Myhrvold and Peter Rinearson)

Business @ the Speed of Thought (with Collins Hemingway)

How to Avoid a Climate Disaster

HOW TO PREVENT THE NEXT PANDEMIC

HOW TO PREVENT THE NEXT PANDEMIC

BILL GATES

ALFRED A. KNOPF NEW YORK TORONTO 2022

THIS IS A BORZOI BOOK PUBLISHED BY
ALFRED A. KNOPF AND ALFRED A. KNOPF CANADA

www.aaknopf.com
www.penguinrandomhouse.ca

Knopf, Borzoi Books, and the colophon are registered trademarks of Penguin Random House LLC. Knopf Canada and colophon are trademarks of Penguin Random House Canada Limited.

Library of Congress Cataloging-in-Publication Data
Names: Gates, Bill, 1955– author.
Title: How to prevent the next pandemic / Bill Gates.
Description: First edition. | New York : Alfred A. Knopf, 2022. |
Includes bibliographical references and index.
Identifiers: LCCN 2021062526 | ISBN 9780593534489 (hardcover) |
ISBN 9780593534496 (ebook)
Subjects: MESH: Pandemics—prevention & control |
COVID-19—prevention & control | Popular Work
Classification: LCC RA644.C67 | NLM WA 105 |
DDC 614.5/92414—dc23/eng/20220119
LC record available at https://lccn.loc.gov/2021062526

Library and Archives Canada Cataloguing in Publication
Title: How to prevent the next pandemic / Bill Gates.
Names: Gates, Bill, 1955– author.
Description: Includes bibliographical references and index.
Identifiers: Canadiana (print) 20220142092 | Canadiana (ebook) 2022014219X |
ISBN 9781039005020 (hardcover) | ISBN 9781039005037 (EPUB)
Subjects: LCSH: Pandemics—Prevention—Popular works. |
LCSH: COVID-19 Pandemic, 2020– —Popular works.
Classification: LCC RA643 .G38 2022 |
DDC 362.1028/9—dc23

Jacket design by Carl De Torres

Manufactured in the United States of America
First Edition

To the frontline workers who risked their lives during
COVID, and to the scientists and leaders who can
make sure they never have to do it again

And in memory of Dr. Paul Farmer, who inspired
the world with his commitment to saving lives.
Author proceeds from this book will be donated
to his organization, Partners in Health.

CONTENTS

HOW TO PREVENT THE NEXT PANDEMIC

INTRODUCTION

I was having dinner on a Friday night in mid-February 2020 when I realized that COVID-19 would become a global disaster.

For several weeks, I had been talking with experts at the Gates Foundation about a new respiratory disease that was circulating in China and had just begun to spread elsewhere. We're lucky to have a team of world-class people with decades of experience in tracking, treating, and preventing infectious diseases, and they were following COVID-19 closely. The virus had begun to emerge in Africa, and based on the foundation's early assessment and requests from African governments, we had made some grants to help keep it from spreading further and to help countries prepare in case it took off. Our thinking was: We hope this virus won't go global, but we have to assume it will until we know otherwise.

At that point, there was still reason to hope that the virus could be contained and wouldn't become a pandemic. The Chinese government had taken unprecedented safety measures to lock down Wuhan, the city where the virus emerged—schools and public places were closed, and citizens were issued permission cards that allowed them to leave their homes every other day for thirty minutes at a time. And the virus was still limited enough that countries were letting people travel freely. I had flown to South Africa earlier in February for a charity tennis match.

When I got back from South Africa, I wanted to have an in-depth conversation about COVID-19 at the foundation. There was one central question I could not stop thinking about and wanted to explore at length: Could it be contained, or would it go global?

I turned to a favorite tactic that I've been relying on for years: the working dinner. You don't bother with an agenda; you simply invite a dozen or so smart people, provide the food and drinks, tee up a few questions, and let them start thinking out loud. I've had some of the best conversations of my working life with a fork in my hand and a napkin in my lap.

So a couple of days after returning from South Africa, I sent an email about scheduling something for the coming Friday night: "We could try and do a dinner with the people involved with coronavirus work to touch base." Almost everyone was nice enough to say yes—despite the timing and their busy schedules—and that Friday, a dozen experts from the foundation and other organizations came to my office outside Seattle for dinner. Over short ribs and salads, we turned to that key question: Would COVID-19 turn into a pandemic?

As I learned that night, the numbers were not in humanity's favor. Especially because COVID-19 spread through the air—making it more transmissible than, say, a virus that is spread through contact, like HIV or Ebola—there was little chance of containing it to a few countries. Within months, millions of people all over the world were going to contract this disease, and millions would die from it.

I was struck that governments weren't more concerned about this looming disaster. I asked, "Why aren't governments acting more urgently?"

One scientist on the team, a South African researcher named

Keith Klugman, who came to our foundation from Emory University, simply said: "They should be."

Infectious diseases—both the kind that turn into pandemics and the kind that don't—are something of an obsession for me. Unlike the subjects of my previous books, software and climate change, deadly infectious diseases are not generally something that people want to think about. (COVID-19 is the exception that proves the rule.) I've had to learn to temper my enthusiasm for talking about AIDS treatments and a malaria vaccine at parties.

My passion for the subject goes back twenty-five years, to January 1997, when Melinda and I read an article in *The New York Times* by Nicholas Kristof. Nick reported that diarrhea was killing 3.1 million people every year, almost all of them children. We were shocked. Three million kids a year! How could that many children be dying from something that was, as far as we knew, little more than an uncomfortable inconvenience?

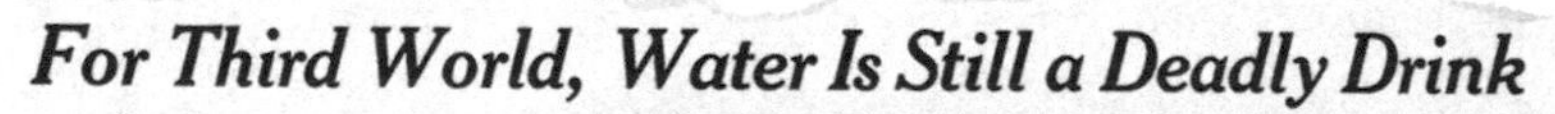

For Third World, Water Is Still a Deadly Drink

By NICHOLAS D. KRISTOF

THANE, India — Children like the Bhagwani boys scamper about barefoot on the narrow muddy ...hs that wind through the laby... ...slum here, squatting and ...mselves as the need ...he filth as

From *The New York Times.* © 1997 The New York Times Company. All rights reserved. Used under license.

We learned that the simple lifesaving treatment for diarrhea—an inexpensive liquid that replaces the nutrients lost during an episode—wasn't reaching millions of children. That seemed like a problem we could help with, and we started making grants to get the treatment out more broadly and to support work on a vaccine that would prevent diarrheal diseases in the first place.*

I wanted to know more. I reached out to Dr. Bill Foege, one of the epidemiologists responsible for the eradication of smallpox and a former head of the Centers for Disease Control and Prevention. Bill gave me a stack of eighty-one textbooks and journal articles on smallpox, malaria, and public health in poor countries; I read them as fast as I could and asked for more. One of the most influential for me had a mundane title: *World Development Report 1993: Investing in Health, Volume 1.* My obsession with infectious diseases—and particularly with infectious diseases in low- and middle-income countries—had begun.

When you start reading up on infectious diseases, it isn't long before you come to the subject of outbreaks, epidemics, and pandemics. The definitions for these terms are less strict than you may think. A good rule of thumb is that an outbreak is when a disease spikes in a local area, an epidemic is when an outbreak spreads more broadly within a country or region, and a pandemic is when an epidemic goes global, affecting more than one continent. And some diseases don't come and go, but stay consistently in a specific location—those are known as *endemic* diseases. Malaria, for instance, is endemic to many equatorial regions. If COVID-19 never goes away completely, it'll be classified as an endemic disease.

It's not at all unusual to discover a new pathogen. In the past fifty years, according to the World Health Organization (WHO), scientists have identified more than 1,500 of them; most began in animals and then spread to humans.

Some never caused much harm; others, such as HIV, have been catastrophes. HIV/AIDS has killed more than 36 million people, and more than 37 million people are living with HIV today. There

* I'll tell you how it turned out in Chapter 3.

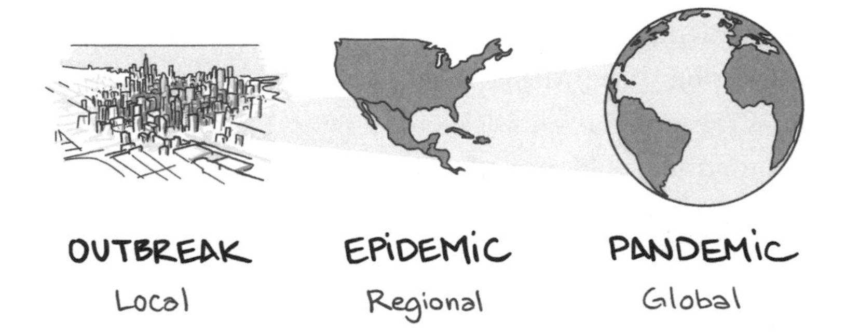

were 1.5 million new cases in 2020, though there are fewer new cases each year because people who are being properly treated with antiviral drugs don't spread the disease.

And with the exception of smallpox—the only human disease ever eradicated—old infectious diseases are still hanging around. Even plague, a disease most of us associate with medieval times, is still with us. It struck Madagascar in 2017, infecting more than 2,400 people and killing more than 200. The WHO receives reports of at least 40 cholera outbreaks every year. Between 1976 and 2018, there were 24 localized outbreaks and one epidemic of Ebola. If you include small ones, there are probably more than 200 outbreaks of infectious diseases every year.

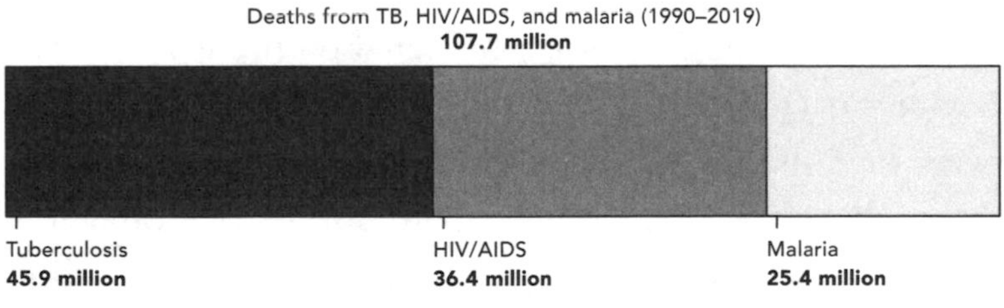

Endemic killers. HIV/AIDS, malaria, and tuberculosis have killed more than 100 million people worldwide since 1990. (Institute for Health Metrics and Evaluation)

AIDS and other "silent epidemics," as they came to be known—tuberculosis, malaria, and others—are the focus of the foundation's global health work, along with diarrheal diseases and maternal mortality. In 2000, these diseases killed more than 15 million people in all, many of them children, and yet shockingly little money was being spent on them. Melinda and I saw this as the area in which our resources and our knowledge of how to build teams to create new innovations could make the biggest difference.

A billboard promoting AIDS awareness and prevention in Lusaka, Zambia.

This is the subject of a common misconception about our foundation's health work. It's not concentrated on protecting people in rich countries from diseases. It's concentrated on the gap in health between high-income countries and low-income ones. Now, in the course of that work, we learn a lot about diseases that can affect the rich world, and some of our funding will help with these diseases, but they are not a focus of our grantmaking. The private sector, rich-country governments, and other philanthropists put a lot of resources into that work.

Pandemics, of course, affect all countries, and I have worried a lot about them since I began studying infectious diseases. Respiratory viruses, including the influenza family and the coronavirus family, are particularly dangerous because they can spread so quickly.

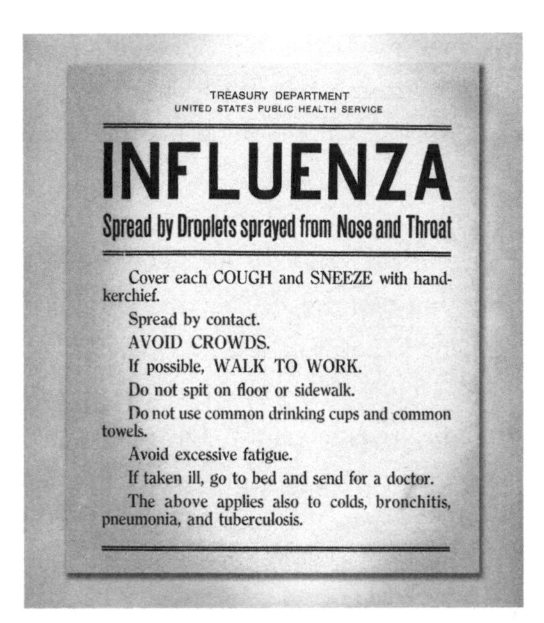

A notice from the U.S. government encouraging proper hygiene and social distancing during the 1918 influenza pandemic.

And the odds that a pandemic will strike are only going up. That's partly because, with urbanization, humans are invading natural habitats at a growing rate, interacting with animals more often, and creating more opportunities for a disease to jump from them to us. It's also because international travel is skyrocketing (or at least it was before COVID slowed its growth): In 2019, before COVID, tourists around the world made 1.4 billion international arrivals every year—up from just 25 million in 1950. The fact that the world had gone a century since a catastrophic pandemic—the most recent one, the flu of 1918, killed something like 50 million people—is largely a matter of luck.

Before COVID, the possibility of a flu pandemic was, relatively speaking, well known; many people had at least heard of the 1918 flu, and they might have remembered the swine flu pandemic of 2009–10. But a century is a long time, so almost no one alive had lived through the flu pandemic, and the swine flu pandemic didn't turn out to be a huge problem because it wasn't much more fatal than the normal flu. At the time I was learning all this, in the early 2000s, coronaviruses—which are one of three virus types that cause most common colds—weren't discussed nearly as often as the flu.

The more I learned, the more I realized just how unprepared the world was for a serious respiratory virus epidemic. I read a report on the WHO's response to the 2009 swine flu pandemic that concluded, prophetically: "The world is ill-prepared to respond to a severe influenza pandemic or to any similarly global, sustained and threatening public-health emergency." The report laid out a step-by-step plan for getting prepared. Few of the steps were taken.

The next year, my friend Nathan Myhrvold started telling me about some research he was doing into the greatest threats faced by humanity. Although his biggest worry was an engineered bioweapon—a disease made in a lab—naturally occurring viruses were high on the list.

I've known Nathan for decades: He created Microsoft's cutting-edge research division and is a polymath who has done research on cooking (!), dinosaurs, and astrophysics, among other things. He's not prone to exaggerating risks. So when he argued that governments around the world were doing essentially nothing to prepare for pandemics of any kind, either natural or intentionally created, we talked about how to change that.*

Nathan uses an analogy that I like. Right now, the building you are sitting in (assuming you're not reading this book at the beach) is probably fitted with smoke detectors. Now, the odds that the building you're in will burn down today are very low—in fact, it might go 100 years without burning down. But that building isn't the only one around, of course, and somewhere in the world, at this very

* Nathan eventually wrote a paper about these ideas called "Strategic Terrorism: A Call to Action" for the journal *Lawfare*. You can find it at https://papers.ssrn.com. I wouldn't advise reading it before bed—it is sobering.

moment, a building is burning down. That constant reminder is why people install smoke detectors: to protect against something that's rare but potentially very destructive.

When it comes to pandemics, the world is one big building fitted with smoke detectors that aren't especially sensitive and have trouble communicating with one another. If there's a fire in the kitchen, it might spread to the dining room before enough people hear about it to go put it out. Plus, the alarm only goes off about every 100 years, so it is easy to forget that the risk is there.

It's hard to get your head around just how quickly a disease can spread, because exponential growth isn't something that most of us encounter in our day-to-day lives. But consider the math. If 100 people have an infectious disease on Day 1, and if the number of cases doubles every day, the entire population of the earth will be infected by Day 27.

In the spring of 2014, I started getting emails from the health team at the foundation about an outbreak that sounded ominous: A few cases of Ebola virus had been identified in southeastern Guinea. By that July, Ebola cases had been diagnosed in Conakry, the capital of Guinea, and in the capital cities of Guinea's neighbors, Liberia and Sierra Leone. Eventually the virus would spread to seven other countries, including the United States, and more than 11,000 people would die.

Ebola is a scary disease—it frequently causes patients to bleed from their orifices—but its rapid onset and immobilizing symptoms mean that it can't infect tens of millions of people. Ebola spreads only through physical contact with the bodily fluids of an infected person, and by the time you're really infectious, you're too sick to move around. The biggest risks were to people who were taking care of Ebola patients, either at home or in the hospital, and during funeral rites, when someone would wash the body of a person who had died of the disease.

Even though Ebola wasn't going to kill many Americans, it did

During the Ebola epidemic in West Africa of 2014–16, many people contracted the virus during funeral gatherings because they came into close contact with a recent victim of the disease.

remind them that an infectious disease can travel long distances. In the Ebola outbreak, a frightening pathogen had come to the United States as well as the United Kingdom and Italy—places that American tourists liked to visit. The fact that there had been a total of six cases and one death in those three countries, versus more than 11,000 in West Africa, didn't matter. Americans were paying attention to epidemics, at least for the moment.

I thought it might be an opportunity to highlight the fact that the world wasn't ready to handle an infectious disease that really could cause a pandemic. *If you think Ebola is bad, let me tell you what the flu could do.* Over the Christmas holidays of 2014, I started writing a memo about the gaps in the world's readiness that had been highlighted by Ebola.

They were enormous. There was no systematic way to monitor the progress of disease through communities. Diagnostic tests, when they were available, took days to return results—an eternity when you need to isolate people if they're infected. There was a volunteer network of brave infectious-disease experts who went to help authorities in the affected countries, but there wasn't a large full-time team of paid experts. And even if there had been such a team, there was no plan in place to move them to where they needed to be.

In other words, the problem was not that there was some system in place that didn't work well enough. The problem was that there was hardly any system at all.

I still didn't think it made sense for the Gates Foundation to make this one of its top priorities. After all, we focus on areas where the markets fail to solve big problems, and I thought that the governments of rich countries would get in gear after the Ebola scare, if they understood what was at stake. In 2015, I published a paper in *The New England Journal of Medicine,* pointing out how unprepared the world was and laying out what it would take to get ready. I adapted the warning for a TED talk called "The Next Epidemic? We're Not Ready," complete with an animation showing 30 million people dying from a flu as infectious as the 1918 one. I wanted to be alarming to make sure the world got ready—I pointed out that there would be trillions of dollars of economic losses and massive disruption. This TED talk has been viewed 43 million times, but 95 percent of those views have come since the COVID pandemic started.

The Gates Foundation, in partnership with the governments of Germany, Japan, and Norway, and the Wellcome Trust, created an organization called CEPI—the Coalition for Epidemic Preparedness Innovations—to accelerate work on vaccines against new infectious diseases and help those vaccines reach people in the poorest countries. I also funded a local study in Seattle to learn more about how the flu and other respiratory diseases move through a community.

Although CEPI and the Seattle Flu Study were good investments that helped when COVID came, not much else was accomplished. More than 110 countries analyzed their preparedness and the WHO outlined steps to close the gaps, but nobody acted on these assessments and plans. Improvements were called for but never made.

Six years after I gave my TED talk and published that *NEJM* paper, as COVID-19 was spreading around the world, reporters and friends would ask me if I wished I had done more back in 2015. I don't know how I could have gotten more attention on the need for better tools and practice scaling them up rapidly. Maybe I should have written this book in 2015, but I doubt many people would have read it.

In early January 2020, the Gates Foundation team we had set up to monitor outbreaks after the Ebola scare was tracking the spread of SARS-CoV-2, the virus we now know as the one that causes COVID-19.*

On January 23, Trevor Mundel, who leads our global health work, sent Melinda and me an email outlining his team's thinking and requesting the first round of funding for COVID work.

* A word about terminology. SARS-CoV-2 is the name of the virus that causes the disease COVID-19. Technically, *COVID* refers to all diseases caused by coronaviruses, of which COVID-19 is one. (The 19 comes from the fact that it was discovered in 2019.) But to keep things readable, from here on out, I'm going to use *COVID* to refer to both the disease COVID-19 and the virus that causes it.

"Unfortunately," he wrote, "the coronavirus outbreak continues to spread with the potential to become a serious pandemic (too early to be sure but essential to act now)."*

Melinda and I have long had a system for making decisions about urgent requests that can't wait for our annual strategy reviews. Whoever sees it first sends it to the other and says, basically, "This looks good, do you want to go ahead and approve it?" Then the other one sends an email approving the spending. As co-chairs, we still use this system for making big decisions related to the foundation, even though we're no longer married and are now working with a board of trustees.

Ten minutes after Trevor's mail came through, I suggested to Melinda that we approve it; she agreed and replied to Trevor: "We are approving $5M [i.e., $5 million] today and realize there may be an additional amount needed in the future. Glad the team has jumped on top of this so quickly. It is very concerning."

As both of us suspected, there were definitely additional amounts needed, as became clear at the mid-February dinner and many other meetings. The foundation has committed more than $2 billion to various aspects of fighting COVID, including slowing its spread, developing vaccines and treatments, and helping make sure that these lifesaving tools reach people in poor countries.

Since the pandemic began, I've had the chance to work with and learn from countless health experts at the foundation and outside it. One deserves special mention.

In March 2020, I had my first call with Anthony Fauci, the head of the infectious diseases institute of the National Institutes of Health. I'm lucky to have known Tony for years (long before he was

* I've already mentioned the Gates Foundation several times in this Introduction, and I'll be mentioning it more throughout this book. This is not because I want to brag, but because the foundation's teams played an important part in much of the effort to develop vaccines, treatments, and diagnostics for COVID-19. It would be hard to tell that story without mentioning their work.

on the cover of pop-culture magazines), and I wanted to hear what he was thinking about all this—especially the potential for various vaccines and treatments that were being developed. Our foundation was backing many of them, and I wanted to make sure our agenda for developing and deploying innovations was aligned with his. Also, I wanted to understand what he was saying publicly about things like social distancing and wearing masks so I could help by echoing the same points when I did interviews.

We had a productive first call, and Tony and I would check in monthly for the rest of the year, discussing the progress on different treatments and vaccines and strategizing about how work done in the United States could benefit the rest of the world. We even did a few interviews together. It was an honor to sit next to him (virtually, of course).

One side effect of speaking out, though, is that it has provoked more of the criticisms of the Gates Foundation's work that I've been hearing for years. The most thoughtful version goes like this: Bill Gates is an unelected billionaire—who is he to set the agenda on health or anything else? Three corollaries of this criticism are that the Gates Foundation has too much influence, that I have too much faith in the private sector as an engine of change, and that I'm a technophile who thinks new inventions will solve all our problems.

It's certainly true that I've never been elected to any public position, and I don't plan to seek one. And I agree that it's not good for society when rich people have undue influence.

But the Gates Foundation does not use its resources or its influence in secret. We're open about what we fund and what the results have been—the failures as well as the successes. And we know that some of our critics don't speak up because they don't want to risk losing grants from us, which is one of the reasons we make extra efforts to consult outside experts and seek out different viewpoints. (We expanded our board of trustees in 2022 for similar reasons.) We aim to improve the quality of the ideas that go into public policies and to steer funding toward those ideas that are likely to have the greatest impact.

Critics are also correct that the foundation has become a very large funder of some big initiatives and institutions that are predominantly the preserve of governments, such as the fight against polio and support for organizations like the WHO. But this is largely because these are areas of great need that do not get nearly enough funding and support from governments even though, as this pandemic has shown, they clearly benefit society as a whole. Nobody would be happier than I would if the Gates Foundation's funding became a much smaller proportion of global spending in the coming years—because, as this book will argue, these are investments in a healthier, more productive world.

On a related point, critics argue that it's not fair that a few people like me got wealthier during the pandemic, while so many other people suffered. They're absolutely right. My wealth has largely insulated me from the impact of COVID—I do not know what it is like to have your life devastated by this pandemic. The best I can do is to uphold the pledge I made years ago to return the vast majority of my resources to society in ways that make the world a fairer place.

And yes, I am a technophile. Innovation is my hammer, and I try to use it on every nail I see. As a founder of a successful technology company, I am a great believer in the power of the private sector to drive innovation. But innovation doesn't have to be just a new machine or a vaccine, as important as those are. It can be a different way of doing things, a new policy, or a clever scheme for financing a public good. In this book you'll read about some of these innovations, because great new products only do the most good if they reach the people who need them most—and in health, that often requires working with governments, which even in the poorest countries are nearly always the entities that provide public services. That's why I'll make the case for strengthening public health systems, which—when functioning well—can serve as the first line of defense against emerging diseases.

Unfortunately, not every criticism of me is as thoughtful. Throughout COVID, I've marveled at how I became the target of wild conspiracy theories. It's not an entirely new sensation—nutty ideas about Microsoft have been around for decades—but the attacks are more intense now. I have never known whether to engage with them or not. If I ignore them, they keep spreading. But does it actually persuade anyone who buys into these ideas if I go out and say, "I am not interested in tracking your movements, I honestly don't care where you're going, and there is no movement tracker in any vaccine"? I've decided that the best way forward is to just keep doing the work and believe that the truth will outlive the lies.

Years ago, the eminent epidemiologist Dr. Larry Brilliant coined a memorable phrase: "Outbreaks are inevitable, but pandemics are optional." Diseases have always spread among humans, but they don't have to become global disasters. This book is about how governments, scientists, companies, and individuals can build a system that will contain the inevitable outbreaks so they don't become pandemics.

There is, for obvious reasons, more momentum than ever to do this now. Nobody who has been through COVID will ever forget it. Just as World War II changed the way my parents' generation looked at the world, COVID has changed the way we see the world.

But we do not have to live in fear of another pandemic. The world can provide basic care to everyone, and be ready to respond to and contain any emerging diseases.

What would it look like in practice? Imagine:

> Research allows us to understand all respiratory pathogens and prepares us to create tools like diagnostics, antiviral drugs,

and vaccines at higher volumes and far faster than is possible today.

Universal vaccines protect everyone from every strain of the respiratory pathogens most likely to cause a pandemic—coronaviruses and influenza.

A potentially threatening disease is rapidly detected by local public health agencies, which function effectively in even the world's poorest countries.

Anything out of the ordinary is shared with capable labs for study, and the information is uploaded to a global database monitored by a dedicated team.

When a threat is detected, governments sound the alarm and initiate public recommendations for travel, social distancing, and emergency planning.

Governments start using the blunt tools that are already on hand, such as mandatory quarantines, antivirals that protect against almost any strain, and tests that can be performed in any health clinic, workplace, or home.

If that isn't sufficient, then the world's innovators immediately get to work developing tests, treatments, and vaccines for the pathogen. Diagnostics in particular ramp up extremely fast so that large numbers of people can be tested in a short time.

New drugs and vaccines are approved quickly, because we've agreed ahead of time on how to run trials quickly and share the results. Once they're ready to go into production, manufacturing gears up right away because factories are already in place and approved.

No one gets left behind, because we've already worked out how to rapidly make enough vaccines for everyone.

Everything gets where it's supposed to, when it's supposed to, because we've set up systems to get products delivered all the way to the patient. Communications about the situation are clear and avoid panic.

And this all happens quickly. It takes just six months to go from raising the first red flag to making enough safe, effective* vaccines to protect the population of the earth.

To some people reading this book, the scenario I've just described will sound overly ambitious. It's certainly a big goal, but we're already headed in that direction. In 2021, the White House announced a plan for developing a vaccine in the next epidemic within 100 days, if resources are allocated. And lead times are already shrinking: It took just twelve months from the time the COVID virus was analyzed genetically until the moment the first vaccines were tested and ready for use, a process that usually takes at least half a decade. And technological advances made during this pandemic will speed things up even more in the future. If we—governments, funders, private industry—make the right choices and investments, we can do this. In fact, I see an opportunity not just to prevent bad things from happening, but to accomplish something extraordinary: eradicating

* In the medical field, *effectiveness* and *efficacy* mean different things. Efficacy is a measure of how well a vaccine works in a clinical trial. Effectiveness is a measure of how well it works in the real world. For the sake of simplicity, and because *efficacious* is an eyesore of a word, I'll use *effectiveness* to mean both.

entire families of respiratory viruses. That would mean the end of coronaviruses like COVID—and even the end of the flu. Every year, influenza alone causes around one billion illnesses, including 3 million to 5 million severe cases where someone ends up in the hospital. And it kills at least 300,000 people. Add in the impact of coronaviruses, some of which cause the common cold, and the benefits of eradication would be staggering.

Each chapter in this book explains one of the steps we need to take to get ready. Together, they add up to a plan for eliminating the pandemic as a threat to humanity and reducing the chance that anyone ever has to live through another COVID.

One final thought before we dive in: COVID is a fast-moving disease. In the time since I started writing this book, several variants of the virus have appeared, most recently Omicron, and others have vanished. Some treatments that appeared very promising in early studies turned out to be less effective than some people (including me) had hoped. There are questions about vaccines, including how long they provide protection, that can only be answered over time.

In this book I have done my best to write what is true at the time of publication, with the understanding that the state of play will inevitably change in the coming months and years. In any case, the key points of the pandemic prevention plan that I propose will be relevant just the same. Regardless of what COVID does, the world still has a lot of work to do before it can hope to keep outbreaks from turning into global disasters.

CHAPTER 1

LEARN FROM COVID

It's easy to say that people never learn from the past. But sometimes we do. Why hasn't there been a World War III yet? Partly because, in 1945, world leaders looked at history and decided there were better ways to settle their differences.

That's the spirit in which I look at the lessons from COVID. We can learn from it and decide to do a better job of protecting ourselves from deadly diseases—in fact, it's imperative to put a plan in place and fund it now, before COVID becomes yesterday's news, the sense of urgency fades, and the world's attention moves on to something else.*

Many reports have documented the good and the bad of the world's response to COVID, and I've learned a lot from them. I have also pulled together a number of key lessons from my work in global health, including projects such as polio eradication, and from following the pandemic day to day with experts at the foundation

* About the word *we:* I use it in various ways in this book. Sometimes I'm referring to work I am personally involved in (or the Gates Foundation is). But for the sake of simplicity, I also use *we* to refer to the global health sector more broadly, or to the world at large. I will try to make my meaning clear in context.

and in governments, academia, and the private sector. A key element is to look at the countries that did better than others.

Doing the right things early pays huge dividends later.

I know this sounds odd, but my favorite website is a treasure trove of data that tracks diseases and health problems all over the world. It's called the Global Burden of Disease,* and the level of detail it contains is astonishing. (The 2019 version tracked 286 causes of death and 369 types of diseases and injuries in 204 countries and territories.) If you're interested in how long people live, what makes them sick, and how these things change over time, this site is the best source. I can spend hours at a time looking at the data.

The site is published by the Institute for Health Metrics and Evaluation (IHME), which is located at the University of Washington in my hometown of Seattle. As you can probably guess from its name, IHME specializes in measuring health around the world. It also does computer modeling that attempts to establish cause-and-effect relationships: Which factors might explain why cases are going up or down in some country, and what does the forecast look like?

Since early 2020, I've been peppering the team at IHME with questions about COVID. What I've hoped to find out is what the countries that are dealing most successfully with COVID have in common. What did they all do right? Once we answer that question with some certainty, we'll understand the best practices and be able to encourage other countries to adopt them.

The first thing you have to do is define success, but that's not

* https://vizhub.healthdata.org/gbd-compare/

as easy as you might think. You can't just look at how often people with COVID in a given country went on to die from it. That statistic will be skewed by the fact that older people are more likely to die from COVID than younger people, so countries with especially old populations will almost inevitably look worse. (One country that did particularly well—even though it has the world's oldest population—is Japan. It had the best compliance with mask mandates of any country, which helps explain some of its success, but other factors were probably also at play.)

What you really look for in a measure of success is a number that captures the overall impact of the disease. People who die of heart attacks because the hospital is too overwhelmed by COVID patients to treat them ought to be counted just as much as people who die of the disease itself.

There's a measure that does exactly that: It's called excess mortality, and it includes people who die because of the disease's ripple effect as well as those who die directly from COVID. (It's the number of excess deaths per capita, in order to account for the size of a country's population.) The lower your excess mortality, the better you're doing. In fact, some countries' excess mortality is actually negative. That's because they had relatively few deaths from COVID, and there were also fewer traffic accidents and other fatal incidents because people were staying home so much more.

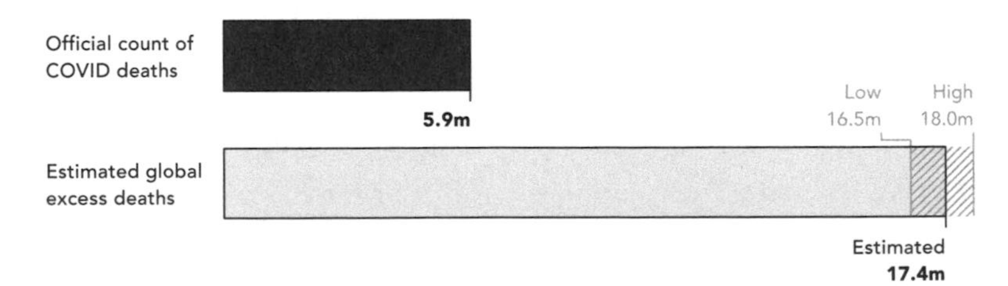

The true toll of COVID. "Excess deaths" measures the impact of COVID by including people whose deaths were indirectly caused by the pandemic. The top bar shows the number of COVID deaths through December 2021. The bottom bar shows the estimated number of excess deaths, with a range between 16.5 million and 18 million. (IHME)

Toward the end of 2021, America's excess mortality was more than 3,200 per million people, roughly on par with Brazil's and Iran's. Canada's, by contrast, was around 650, while Russia's was well over 7,000.

Many of the countries with the lowest excess mortality (near zero or negative)—Australia, Vietnam, New Zealand, South Korea—did three things well early in the pandemic. They tested a large share of the population quickly, isolated people who tested positive or had been exposed, and carried out a plan for detecting, tracing, and managing cases that may have come across their borders.

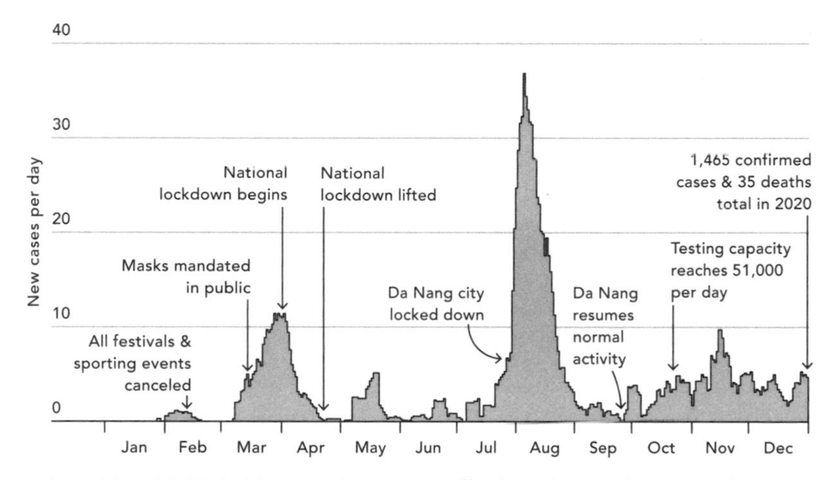

Containing COVID in Vietnam. Government officials implemented measures for controlling the virus during 2020. Having just thirty-five deaths over an entire year in a country of 97 million people is a major accomplishment. (Exemplars in Global Health program)

Unfortunately, early success can be hard to maintain. Relatively few people in Vietnam were vaccinated for COVID—partly because of the limited supply of vaccines, and partly because vaccines didn't seem as urgent when the country had done such a good job controlling the virus. So when the much more transmissible Delta variant came along, there were relatively few people in Vietnam who had any immunity, and the country was hit hard. Its rate of excess deaths went from just over 500 per million people in July 2021 to nearly 1,500 per million people in December—though even at the higher rate, Vietnam was still doing better than the United States. Overall, it was better off having taken those early measures.

IHME's data also suggests that a country's success against COVID correlates roughly with how much people there trust the government. This makes intuitive sense, since if you have confidence in your government, you're more likely to follow its guidelines for preventing COVID. On the other hand, trust in government is measured by polls, and if you live under an especially repressive regime, you're probably not going to tell a random pollster what you really think about the government. And in any case, this finding doesn't easily translate into practical advice that can be implemented quickly. Building trust between people and their government takes years of painstaking, purposeful work.

Another approach to identifying what works is to look at the problem from the other end: Find exemplars that did individual things especially well and study how they did them so that others can do the same. A group called, appropriately enough, Exemplars in Global Health is doing just that, and they have made some fascinating connections.

For example, all other things being equal, countries whose health systems function well in general were more likely to respond well to COVID. If you have a strong network of health clinics that are well staffed with trained personnel, are trusted by people in their community, have supplies when they're needed, and so on, you are in a better position to fight off a new disease. This suggests that any pandemic prevention plan needs to include, among other things, helping low- and middle-income countries improve their health systems. We'll return to this subject in Chapters 8 and 9.

Another example: The data suggests that cross-border trucking was responsible for a fair amount of spread from one country to another. So which places managed it well? Early in the pandemic, Uganda required COVID tests for all truckers coming into the country, and the region of East Africa followed suit soon after. But because the testing process was slow and kits were in short supply, the policy caused major backups at the border—of up to four days—and transmission went up while truckers waited around in cramped quarters.

Uganda and its neighbors did several things to fix the logjam, including dispatching mobile testing labs to border crossings, creating an electronic system to track and share results, and requiring truckers to get tested in the country where they started their route, rather than at the border. Soon, traffic was flowing again, and cases were kept under control.

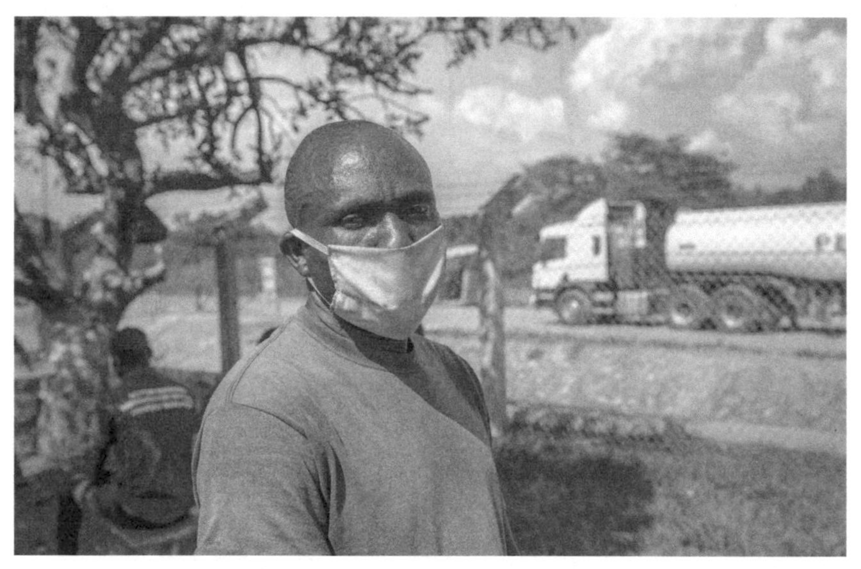

Truck driver Naliku Musa waits for the results of his COVID test at the border between Uganda and South Sudan.

Bottom line: In the early days, if you're able to test a large share of your population, isolate the positive cases and their contacts, and handle possible cases coming from abroad, you'll be well positioned to keep the caseload manageable. If you don't do those things quickly, then only extreme measures can prevent a large number of infections and deaths.

Some countries show us what not to do.

I don't like to dwell on failures, but some are too egregious to ignore. Although there are positive exemplars, most countries handled at least some aspects of their COVID response poorly. I'm calling out the United States here because I know its situation well, and because it should have done so much better than it did, but by no means is it the only country that made a lot of mistakes.

The White House's response in 2020 was disastrous. The president and his senior aides downplayed the pandemic and gave the public terrible advice. Incredibly, federal agencies refused to share data with one another.

It certainly didn't help that the director of the Centers for Disease Control and Prevention is a political appointee subject to political pressure, and some of the CDC's public guidance was clearly influenced by politics. Even worse, the person running the CDC in 2020 wasn't trained as an epidemiologist. The former CDC directors who are still remembered today for their amazing work—people like Bill Foege and Tom Frieden—were experts who had spent much or all of their careers in the organization. Imagine a general who has never even been through a battle simulation suddenly having to run a war.

One of the worst failures, though, is that the United States never got testing right: Not nearly enough people were tested, and results took far too long to come back. If you're carrying the virus but don't know it for another seven days, you've just spent a week potentially infecting other people. To me, the most mind-boggling problem—because it would have been so easy to avoid—is that the U.S. government never fully maximized the capacity for testing people, nor did it create a centralized way to both identify those who should be first in line to get quick results and record the outcomes of all tests. Even two years into the pandemic, as Omicron spread rapidly, many people weren't able to get tested, even when they had symptoms.

In the early months of 2020, any people in America who were worried about having COVID should have been able to go to a government website, answer a few questions about symptoms and risk factors (such as age and location), and find out where they could get tested. Or, if test supplies were limited, the site might determine that their case wasn't a high enough priority and notify them when they *could* be tested.

Not only would the site have made sure that testing kits were used most efficiently—for the people most likely to actually test positive—it also would have given the government additional information about parts of the country where too few people were showing an interest in getting tested. With this data, the government could have directed more resources toward getting the word out and expanding testing in those areas. The site would also have provided people with instant eligibility to participate in a clinical trial if they tested positive or were at high risk, and it could later have been used to help make sure that vaccines went to the people at the highest risk of getting severely sick or dying. And the site would also be useful in nonpandemic times for fighting other infectious diseases.

Any software company worth its salt could have built this site in no time,* but instead states and cities were left to their own devices, and the whole process was chaotic. It was like the Wild West. I remember one especially heated call with people from the White House and CDC in which I was quite rude about their refusal to take this basic step. To this day I don't understand why they wouldn't

* Microsoft would've done it for free, and I'm sure many other companies would have too.

let the most innovative country in the world use modern communications technology to fight a deadly disease.

In the face of something the world should have been better prepared for, people did heroic work.

Whenever there's a disaster, the children's TV host Fred Rogers used to say, "Look for the helpers. You will always find people who are helping." During COVID, it takes very little looking to find the helpers. They are everywhere, and I've had the pleasure of meeting some of them and learning about many more.

Every day for five months of 2020, as a COVID tester in Bengaluru, India, Shilpashree A.S. would put on a protective gown, goggles, latex gloves, and a mask. (Like many people in India, she uses initials referring to her hometown and her father's name as her

Shilpashree A.S. takes samples in Bengaluru, India, while stationed in a booth and wearing protective gear.

last name.) Then she'd step into a tiny booth with two holes for her arms and spend hours performing nasal swab tests on long lines of patients. To protect her family, she had no physical contact with them—for five months they saw one another only on video calls.

Thabang Seleke was one of 2,000 volunteers in Soweto, South Africa, to participate in a study on the effectiveness of the COVID vaccine developed at Oxford University. The stakes for his country were high: By September 2020, more than 600,000 people had been diagnosed with COVID and more than 13,000 people had died from it. Thabang heard about the trial from a friend and stepped forward to help bring an end to the coronavirus in Africa and beyond.

Sikander Bizenjo went from Karachi to his home province of Balochistan, a dry, mountainous region in southwestern Pakistan where 70 percent of the population lives in poverty. He founded a group called Balochistan Youth Against Corona, which has trained more than 150 young boys and girls to help people across the province. They're hosting COVID awareness sessions in local languages while also building reading rooms and donating hundreds of thousands of books. They've provided medical equipment to 7,000 families and food to 18,000 families.

Ethel Branch, a member of the Navajo Nation and its former attorney general, left her law firm to help form the Navajo & Hopi Families COVID-19 Relief Fund, an organization that delivers water, food, and other necessities to people in need throughout the Navajo and Hopi nations. She and her colleagues have raised millions of dollars (some of it through one of the top five GoFundMe campaigns of 2020) and organized hundreds of young volunteers who have helped tens of thousands of families from both nations.

The stories of people who are making sacrifices to help others during this crisis could fill an entire book. Around the world, health care workers put themselves at risk to treat sick people—according to the WHO, more than 115,000 had lost their lives taking care of COVID patients by May 2021. First responders and frontline workers kept showing up and doing their jobs. People checked in on neighbors and bought groceries for them when they couldn't leave home. Countless people followed the mask mandates and stayed home as much as possible. Scientists worked around the clock, using all their brainpower to stop the virus and save lives. Politicians made decisions based on data and evidence, even though these decisions weren't always the popular choice.

Not everyone did the right thing, of course. Some people have refused to wear masks or get vaccinated. Some politicians have denied the severity of the disease, shut down attempts to limit its spread, and even implied that there's something sinister in the vaccines. It's impossible to ignore the impact their choices are having on millions of people, and there's no better proof of those old political clichés: Elections have consequences, and leadership matters.

Expect variants, surges, and breakthrough cases.

Unless you work on infectious diseases, you had probably never heard of variants until COVID. The idea may have seemed new and scary, but there's nothing particularly unusual about variants. Influenza viruses, for instance, can quickly mutate into new variants—which is why flu vaccines are reviewed each year and frequently updated. Variants of concern are the ones that are more transmissible than others, or better at evading the human immune system.

Early in the pandemic, there was a broad belief in the scientific community that, although there would be some mutations of COVID, they wouldn't cause a big problem. By early 2021, scientists knew that variants were emerging, but they appeared to be evolving in similar ways, leading some scientists to hope that the world had already seen the worst mutations that the virus was capable of. But the Delta variant proved otherwise—its genome had evolved to make it far more transmissible. The arrival of Delta was a bad surprise, but it convinced everyone that even more variants could show up. As I finish this book, the world is facing a sweeping wave of Omicron, the fastest-spreading variant to date—and in fact the fastest-spreading virus we've ever seen.

Viral variants are always a possibility. In future outbreaks, scientists will monitor variants closely to make sure that whatever new tools come out will still work on them. But, because every time a virus jumps from one person to another is an opportunity for it to mutate, the most important thing will be to keep doing the things that definitely reduce transmission: Follow the experts' recommendations on masks, social distancing, and vaccines, and make sure low-income countries get vaccines and the other tools they need to fight the pathogen.

Just as the rise of variants wasn't a surprise, neither were so-called breakthrough cases, in which people who have been vaccinated end up getting infected anyway. Until vaccines or drugs can block infections perfectly, some vaccinated people will still become infected. As more people get vaccinated in a given population, the total number of cases will go down, and a growing percentage of the cases that do occur will be breakthroughs.

Here's one way to think about it. Imagine that COVID starts spreading through a town with a fairly low vaccination rate. A thousand people get so sick that they end up in the hospital. Out of those 1,000 severe cases, 10 are breakthroughs.

Then the virus spreads to the next town over, which has a high vaccination rate. That town sees only 100 severe cases, of which 8 are breakthroughs.

In the first town, breakthroughs represented 10 out of 1,000 severe cases, or one percent. In the second, they made up 8 out of 100, or 8 percent of the total. Eight percent sounds like bad news for town #2, right?

But remember, the important number is not the breakthrough

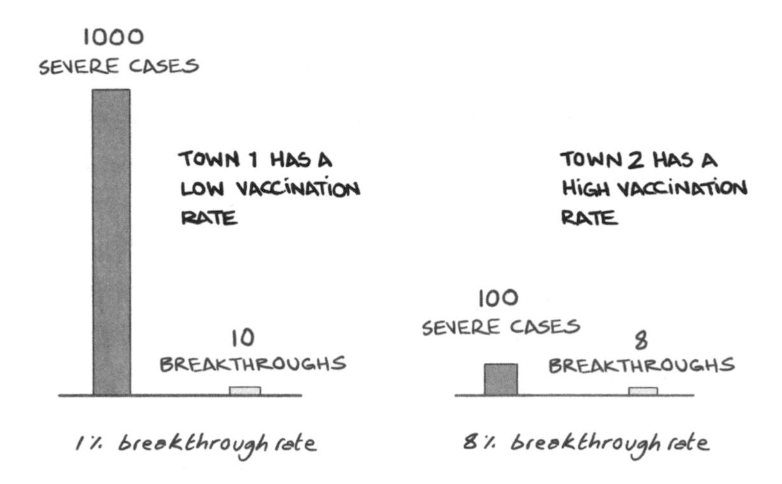

rate. It's the total number of severe cases, and that number went from 1,000 in the first town to just 100 in the next. That is progress by any definition. You're safer by far in town #2, where lots of people are vaccinated, and if you're one of them.

Along with variants and breakthrough cases, waves—big spikes in the number of cases—were not a surprise in and of themselves. We know from the history of previous pandemics that waves occur, yet countries in every region of the world were caught off guard by them. I admit to having been surprised, as many people were, by the size of the Delta wave in India in mid-2021. It was partly the result of wishful thinking—the mistaken idea that the country could relax because it had contained the virus in the early days of 2020. Another explanation is sadly ironic: Countries that do the best job of suppressing the virus early on will often be susceptible to later surges, because their suppression measures kept people from getting sick and developing natural immunity. The aim is to use suppression to delay widespread infection, prevent hospitals from getting overloaded, and buy time for vaccines to protect people. But if an especially transmissible variant shows up before vaccines are widely distributed, and if suppression measures are ended, then a big wave is almost inevitable. India did learn these lessons quite rapidly and ran a successful COVID vaccination campaign later in 2021.

Good science is messy, uncertain, and prone to change.

Here's a partial list of the U.S. government's various positions on wearing masks during COVID:

- Feb. 29, 2020: The surgeon general tweets that people should "STOP BUYING MASKS" because they are "not preventing" COVID (which turned out not to be true) and buying them will make it harder for health workers to access them (which was true at the time, although it was fairly easy to make more masks).
- March 20, 2020: The CDC reiterates that healthy people who do not work in health care or are not taking care of a sick person do not need masks.
- April 3, 2020: Two weeks later, the CDC recommends masks for all people over age two who were in a public setting, traveling, or around others in the same household who might be infected.
- Sept. 15, 2020: The CDC recommends that all teachers and students who attend school in person wear masks whenever feasible.
- Jan. 20, 2021: President Biden signs an executive order requiring masking and physical distancing in federal buildings, on federal lands, and by government contractors. The next day he signs an order requiring masks when

traveling, and nine days later the CDC issues an order that makes refusing to wear a mask in federally mandated spaces a violation of federal law.

- March 8, 2021: The CDC releases new guidance that fully vaccinated people do not need to wear a mask when visiting other vaccinated people inside.
- April 27, 2021: The CDC announces that people do not need to wear masks outdoors when they walk, bike, or run alone or with members of their household regardless of vaccination status. Fully vaccinated people do not need to wear masks outdoors at all unless they're at a large gathering like a concert.
- May 13, 2021: The CDC announces that fully vaccinated people no longer need to wear a mask or physically distance inside. Some states, like Washington and California, continue mask mandates through some or all of June.
- July 27, 2021: The CDC recommends that fully vaccinated people resume wearing masks indoors in parts of the country where case counts are surging. It also recommends that masks be worn indoors by all teachers, staff, students, and visitors to schools regardless of their vaccination status.

You could get whiplash from trying to follow along.

Does this mean the staff of the CDC were incompetent? No. I won't defend every decision the CDC made—as many experts argued at the time, the CDC was wrong in May 2021 when it said that vaccinated people didn't need to wear masks—but during a public health emergency, decisions are made by imperfect people using imperfect data in an environment that's constantly changing. We should have studied respiratory virus transmission a lot more beforehand, instead of having to learn during the pandemic. And expecting perfection during an outbreak actually sets up a perverse dynamic, as the story of David Sencer illustrates.*

Born in Michigan in 1924, Sencer joined the U.S. Navy after graduating from college. After a yearlong bout with tuberculosis, he ended up joining the U.S. Public Health Service, intent on saving people from diseases like the one that had sidelined him for so long.

Sencer made his mark early on with vaccines. After moving to the Centers for Disease Control, he helped draft legislation that created the first broad vaccination program in the U.S. and dramatically expanded the number of children who received the polio vaccine. He became director of the CDC in 1966 and led its expansion into work on malaria, family planning, smoking prevention, and even the quarantine of astronauts after returning from space. Sencer was a master of logistics, a skill that made him indispensable in the successful effort to eradicate smallpox.

In January 1976, a soldier serving at Fort Dix in New Jersey died from swine flu after doing a five-mile march while sick. Thirteen others were hospitalized with the disease. Doctors discovered that the men all had a strain of influenza similar to the one that had caused the 1918 pandemic.

The outbreak never expanded beyond Fort Dix. But in February 1976, worried that there could be a replay of the 1918 disaster when flu season came around that fall—which would mean tens of millions of deaths around the world—Sencer called for mass immunization against this particular strain of swine flu, using an existing vaccine. A presidential panel that included the legendary researchers Jonas Salk and Albert Sabin, both of whom had developed groundbreaking polio vaccines, backed the idea. President Gerald Ford went on television to announce his support for a mass immunization drive, and the campaign quickly kicked into gear.

* Michael Lewis tells Sencer's story well in his book *The Premonition.*

By mid-December, signs of trouble were emerging. Ten states reported cases of vaccinated people contracting Guillain-Barré syndrome, or GBS, an autoimmune disease that causes nerve damage and muscle weakness. The vaccination program was suspended later that month and never reinstated. Shortly thereafter, Sencer was informed that he would be replaced as head of the CDC.

In total, GBS cases occurred in 362 patients out of the 45 million people vaccinated—a rate around four times higher than you would expect with the general population. One study concluded that even if the vaccine did cause GBS in rare cases, its benefits vastly outweighed the risk. But someone needed to take the blame, and Sencer was the fall guy.

Sencer, who died in 2011, remains highly regarded in the world of public health. The consensus is that pushing for mass immunizations was worth the risk: If he had been right about a pandemic, the cost of inaction would have been enormous. But critics focused more on the risk of a rare autoimmune disease—which was real—than on the possibility that tens of millions could die.

In public health, you have to be careful about sending a message that says, "Act early, but you'll be fired if you get it wrong." Of course, if someone makes a truly wretched decision, firing may be in order. But officials need the leeway to make tough calls, because there will always be false alarms, and distinguishing them from the real thing is not an easy task.

What if Sencer had done nothing and his fears had turned out to be valid? Tens of millions of people would have died of a virus that had started in the United States, which had a chance to stop it but chose not to. When people like Sencer act in good faith and with the best data they have available to them, they should not be attacked for possibly having made the wrong call purely because we have the benefit of hindsight. It creates a perverse incentive to be overly cautious—to protect their careers by holding back. And when it comes to public health, holding back can lead to disaster.

It pays to invest in innovation.

It's tempting to assume that invention occurs practically overnight. If in January you wouldn't know what messenger RNA was if it walked up to you on the street, and by July you've read all about it and are getting a vaccine that uses it, you might think it went from idea to reality in just six months. But innovation does not happen in an instant. It takes years of patient, persistent effort by scientists—who will fail more often than they succeed—as well as funding, smart policies, and an entrepreneur's mindset to get an idea out of the lab and into the market.

It's frightening to imagine how much worse COVID would've been if the U.S. government and others hadn't invested years ago in research on vaccines that use messenger RNA (mRNA, which I'll explain in Chapter 6) or another approach called a viral vector. In 2021 alone, they accounted for roughly 6 billion doses delivered worldwide. Without them, we would've been far worse off.

The pandemic has offered up dozens of other concrete examples of innovative ideas, scientific insights, new diagnostic tools, treatments, policies, and even ways to fund the delivery of all these things around the world. Researchers have learned a lot about how viruses move from person to person. And since transmission of the flu virus essentially came to a halt during the first year of COVID, researchers now know that it's possible to stop influenza, which bodes well for future outbreaks of flu and other diseases.

COVID also highlights an inescapable fact about innovation: Most of the world's greatest talent for translating research into commercial products is in the private sector. Not everyone likes that arrangement, but the profit motive is often the most powerful force in the world for getting new products created quickly. It's the government's role to invest in the basic research that leads to major innovations, adopt policies that let new ideas flourish, and create markets and incentives (the way the United States accelerated vaccine work with Operation Warp Speed). And when there are market failures—when the people who most need lifesaving tools can't afford them—then governments, nonprofits, and foundations should step in to fill the gap, often by finding the right way to work with the private sector.

We can do better next time—if we start taking pandemic preparation seriously.

The world responded to COVID faster and more effectively than to any other disease in history. But as the late educator and physician Hans Rosling put it, "Things can be better and bad." In the Better column, for example, I'd put the fact that the world developed safe, effective vaccines in record time. In the Bad column, I'd put the fact that too few people in poor countries are getting them. I'll return to this problem in Chapter 8.

Another entry in the Bad column so far: the world's failure to get serious about preparing for and trying to prevent pandemics.

Governments are responsible for the safety of their people. For common events that cause damage and deaths—fires, natural disasters, wars—governments have a structure for their response: They have experts who understand the risks, get the resources and tools they need, and practice how they'll respond in an emergency. Militaries run large-scale drills to make sure they're prepared for action. Airports run exercises to see if they're ready for an emergency. City, state, and federal governments practice responding to natural disasters. Even schoolchildren go through fire drills and, if they live in the United States, active-shooter drills.

When it comes to pandemics, though, virtually none of this happens. Although people had been raising the alarm for decades about novel diseases that could kill millions of people—a long succession of warnings came before and after mine in 2015—the world didn't respond. For all the effort that humans put into preparing for attacks from fires, storms, and other humans, we had not prepared seriously for an attack by the smallest possible enemy.

In Chapter 2, I'll argue that what we need is a global corps of people whose job is to wake up every day thinking about diseases that could kill huge numbers of people—how to spot them early, how to respond, and how to measure whether we're ready to respond.

To sum up: The world has never invested in the tools it needs or properly prepared for a pandemic. It's time we did. The rest of this book will describe how we can do it.

CHAPTER 2

CREATE A PANDEMIC PREVENTION TEAM

In the year 6 CE, a fire devastated the city of Rome. In its aftermath, the emperor, Augustus, did something that had never been done in the history of the empire: He created a permanent team of firefighters.

The fire brigade, which would grow to include nearly 4,000 men, was equipped with buckets, brooms, and axes and divided into seven groups that stood watch at barracks placed strategically throughout the city. (One of these barracks was discovered in the mid-nineteenth century and is sometimes open to visitors.) Officially, the squad was known as the *Cohortes Vigilum*—which can be loosely translated as "Brothers of the Watch"—but locals came to use the term of endearment *Sparteoli,* or "Little Bucket Fellows."

Elsewhere around the world, China's first professional fire brigade was established in the eleventh century by Emperor Renzong of the Song Dynasty. Europe followed roughly 200 years later. In America, there were volunteer groups before the American Revolution, formed at the urging of a young Benjamin Franklin (who else?), as well as private ones that were paid by insurance companies to save burning buildings. But the United States didn't have a single government-run, full-time firefighting corps until 1853, when one was established by the city of Cincinnati, Ohio.

There are now about 311,000 full-time firefighters in the United States, stationed at nearly 30,000 departments.* Local governments in the U.S. spend more than $50 billion a year staying ready to deal with fires. (I was surprised by how large these numbers are when I looked them up!)

And that's not to mention all the steps we take to prevent fires from starting in the first place. For nearly 800 years, governments have passed laws to reduce the risk of conflagration, including banning thatched roofs (London in the thirteenth century) and requiring the safe storage of fuels for bread ovens (Manchester, England, in the sixteenth). Today, one large fire-prevention nonprofit publishes a list of more than 300 building codes and standards designed to minimize the risk and extent of fires.

In other words, for some 2,000 years, humans have recognized that individual families and businesses aren't solely responsible for protecting themselves—they need help from the community. If your neighbor's house is on fire, your home is at risk, and firefighters will take steps to prevent the flames from spreading. And when they're not actively battling a blaze, the fire department will run drills to keep their skills sharp and help out with other activities related to public safety and service.

Fires don't spread across the entire world, of course, but diseases do. A pandemic is the equivalent of a fire that starts in one building and within weeks is burning in every country in the world. So to prevent pandemics, we need the equivalent of a global fire department.

At the global level, we need a group of experts whose full-time job is to help the world prevent pandemics. It should be responsible for watching out for potential outbreaks, raising the alarm when they emerge, helping to contain them, creating data systems to share case numbers and other information, standardizing policy

* There are also about 740,000 volunteer firefighters in the United States.

recommendations and training, assessing the world's ability to roll out new tools quickly, and organizing drills to look for weak spots in the system. It should also coordinate the many professionals and systems around the globe who do this work at the national level.

Creating this organization requires a serious commitment from rich-country governments, including making sure it is staffed properly. It will be hard to get the right consensus at the global level, as well as the right level of funding—but even knowing the obstacles, I feel it is a critical priority for the world to put this team in place. In this chapter I want to make clear how it should work.

You might think a group like the one I'm proposing already exists. How many movies and TV shows have you seen where there's an outbreak of a scary disease and the world seems perfectly prepared? Someone starts showing symptoms. The president of the United States gets briefed on the situation with a dramatic animated computer model that shows the disease spreading worldwide. A team of experts get the phone call they've been waiting for (always during breakfast with their families, for some reason) and spring into action. Wearing hazmat suits and carrying expensive equipment, they are flown in on helicopters to assess the situation. They take a few samples, speed off to the lab to make the antidote, and go on to save humanity.

Reality is a lot more complicated than that. For one thing, the Hollywood version underplays one of the most important (but admittedly undramatic) tasks in pandemic prevention: making sure that countries have strong health systems. In a well-run system, clinics are fully staffed and equipped, pregnant women get pre- and postnatal care, and children get their routine vaccines; health care workers are well trained in public health and pandemic prevention; and reporting systems make it easy to identify suspicious clusters of cases and raise the red flag. When that kind of infrastructure is in place—as it is in most wealthy nations and some low- and middle-income ones—you're much more likely to notice the early stage of a new disease emerging. Without that infrastructure, you don't notice the new disease until it has spread to tens of thousands of people and probably reached into many countries.

But what's most unrealistic about what you see in the movies is that it suggests there's some agency that pulls together all these different capabilities, acting swiftly and decisively to prevent a pandemic. My favorite example is season 3 of the TV show *24*—a show I really liked—where a terrorist intentionally releases a pathogen in Los Angeles. Word gets to practically every government entity in no time. The hotel where the release took place is immediately sealed off. A computer modeling genius figures out not only how the disease will spread, but how quickly news of the disease will get around, and (the best part) how bad traffic will get as people flee the city. I remember watching those episodes and thinking, "Wow, that government sure knew how to prepare."

It made for great TV, and of course we could all sleep better at night if things really worked that way. But they don't. Although there are many organizations that work hard to respond to a major outbreak, their efforts largely depend on volunteers. (The best known is the Global Outbreak Alert and Response Network, or GOARN.) Regional and national response teams are understaffed and underfunded, and none of them has a mandate from the international community to work globally. The only organization that sort of has that mandate, the WHO, has very little funding and almost no personnel dedicated to pandemics, relying instead on the mostly volunteer GOARN. There is no organization with the size, scope, resources, and responsibility that are essential for detecting and responding to outbreaks and preventing them from becoming pandemics.

Let's consider the sequence of events that are involved in an effective response to an outbreak. Sick people have to go to a clinic, and the health workers there have to diagnose them properly. Those cases must get reported up the chain, and an analyst has to notice an unusual cluster of cases with similar suspicious symptoms or test results. A microbiologist must get samples of the pathogen and determine whether it's something we've seen before. A geneticist may need to map its genome. Epidemiologists have to understand how transmissible and severe the disease is.

Community leaders need to get, and share, accurate information. Quarantines might need to be put in place and enforced. Scientists need to get cracking on diagnostic tests, treatments, and vaccines. And, just as firefighters run drills when they're not putting out a blaze, all of these groups need to have practiced, testing the system to find the weak spots and fix them.

Bits and pieces of what you'd want in a monitor-and-respond system exist. I've met people who have dedicated their lives to this work, and many put their lives on the line for it. But COVID did not happen because there were too few smart, compassionate people trying to prevent it. COVID happened because the world hasn't created an environment in which smart, compassionate people can make the most of their skills as part of a strong, well-prepared system.

What we need is a well-funded global organization with enough full-time experts in all the necessary areas, the credibility and authority that come with being a public institution, and a clear remit to focus on preventing pandemics.

I call it the GERM—Global Epidemic Response and Mobilization—team, and the job of its people should be to wake up every day asking themselves the same questions: "Is the world ready for the next outbreak? What can we do to be better prepared?" They should be fully paid, regularly drilled, and prepared to mount a coordinated response to the next threat of a pandemic. The GERM team should have the ability to declare a pandemic and work with national governments and the World Bank to raise money for the response very quickly.

My back-of-the-napkin estimate is that GERM would need about 3,000 full-time employees. Their skills should run the gamut: epidemiology, genetics, drug and vaccine development, data systems, diplomacy, rapid response, logistics, computer modeling, and communications. GERM should be managed by the World Health Organization, the only group that can give it global credibility, and it should have a diverse workforce, with a decentralized staff working in many places around the world. To get the best staff possible, GERM should have a special personnel system different from what most U.N. agencies have. Most of the team would be based at individual countries' national public health institutes, though some would sit in the WHO's regional offices and at its headquarters in Geneva.

When there's a potential pandemic looming, the world needs expert analysis of early data points that can confirm the threat. GERM's data scientists would build a system for monitoring reports of clusters of suspicious cases. Its epidemiologists would monitor reports from national governments and work with WHO colleagues to identify anything that looks like an outbreak. Its product-development experts would advise governments and companies on the highest-priority drugs and vaccines. GERMers who understand computer modeling would coordinate the work of modelers around the world. And the team would take the lead on creating and coordinating common responses, such as how and when to implement border closures and recommend mask use.

Diplomacy will inevitably be part of the job. After all, national and local leaders are the ones who understand the unique conditions in their country, who speak every local language, who know the key players, and to whom the public looks for leadership. People from GERM would have to work closely with them, making it clear that their job is to support, not supplant, local expertise. If GERM becomes—or even appears to be—something imposed from the outside, some countries will reject its recommendations.

For countries that need additional support, GERM should fund or loan public health experts who would participate in this global pandemic-prevention network. They would train and drill together to keep their skills sharp, and they would stay ready to respond locally or globally when they're needed. Countries with greater need and a high risk of outbreaks would bring in more GERM team members from the network and host them to build local expertise in infectious diseases. Regardless of where they're assigned, these GERMers would have a dual identity: They'd be part of the national detection-and-response system, and also part of GERM's rapid response.

Finally, the GERM team should be responsible for testing the world's monitor-and-respond system to find the weak spots. They would develop a checklist for pandemic preparedness, similar to the ones that airplane pilots follow before every takeoff and many surgeons now go through during an operation. And just as militaries do complex exercises where they simulate different conditions and see how well they respond, the GERM team would organize outbreak response exercises. Not war games, but germ games. This will be the team's most important role, and we'll return to it in much more detail in Chapter 7.

The group I'm describing would be new, but not exactly unprecedented. It's based on a model I've seen work extremely well against another disease, one that we are achingly close to eradicating.

Polio—a paralyzing disease that usually affects the legs but can, in rare cases, affect the diaphragm and make it impossible to breathe—has probably been around for thousands of years. (An Egyptian tablet from the sixteenth century BCE portrays a priest with what appears to be a leg withered by polio.) Even though polio vaccines were invented in the mid-1950s and early 1960s, for decades they didn't reach everyone who needed them. As recently as the late 1980s, there were still 350,000 cases of wild polio every year, in 125 countries.*

But in 1988, the WHO and its partners—led by the volunteer group Rotary International—set out to eradicate polio. By adding a vaccine for it to the list of routine childhood immunizations, and by undertaking massive vaccination campaigns, the world cut cases of wild poliovirus from 350,000 a year to fewer than a dozen in 2021. That's a drop of more than 99.9 percent! And instead of existing in 125 countries, wild polio exists in only two: Afghanistan and Pakistan.

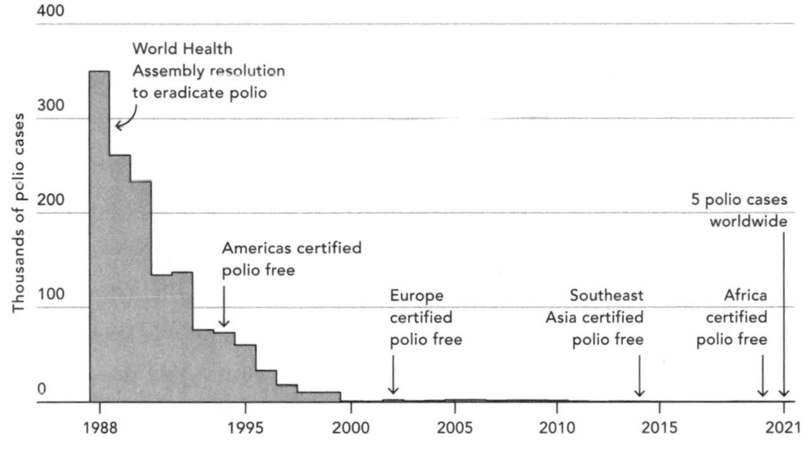

Ending polio. A worldwide effort has caused wild polio cases to plummet, from 350,000 in 1988 to just 5 in 2021. (WHO)

A main ingredient in the secret sauce is what are known as emergency operations centers, or EOCs. These have been around for the past decade, starting in Nigeria and becoming the mainstay of the polio program in more than a dozen countries where polio has been hardest to eliminate.

Picture the office of a political campaign in the last days before

* I'm specifying "wild" poliovirus here to distinguish it from vaccine-derived cases, which are quite rare.

the election, and you'll get an idea of what an EOC looks like. Maps and charts are pasted up on the walls, but instead of tracking poll numbers, they're revealing the latest polio data. It's the nerve center where public health workers from the government and international partners (such as the WHO, UNICEF, CDC, and Rotary International) drive the response to any reports of polio—in a paralyzed child, or if the virus is found in a sewage sample. (I'll explain more about sewage sampling in the next chapter.)

Nigeria's national emergency operations center, in Abuja, is a hub for dealing with public health threats, including Ebola, measles, and Lassa fever—and in 2020, it quickly pivoted to COVID.

EOCs typically oversee the distribution of millions of doses of oral polio vaccine every year, managing tens of thousands of vaccinators who go house to house to vaccinate children multiple times, maintaining relationships with local leaders to stamp out misperceptions and misinformation about vaccines, and using digital tools to find out whether vaccinators are able to reach all of the places they're scheduled to go.

Thanks to this system, the staff at an EOC even know how many households refuse to have their children vaccinated. The measurement is incredibly precise: The coordinator of Pakistan's national EOC reported that they had reduced the refusal rate from 1.7 percent in 2020 to 0.8 percent the next year, and that in one campaign, just 0.3 percent of households had refused the vaccine. And in March 2020, the government used its polio EOC as the model for one focused on COVID.

The GERM team should be a worldwide EOC on steroids. Just as emergency operations centers fight endemic diseases like polio while staying ready to pivot when something new emerges, the GERM team would do double duty as well—only with the focus reversed. Emerging diseases should be their top priority, but when there isn't an active pandemic threat, they would keep their skills sharp by helping out with polio, malaria, and other infectious diseases.

You might have noticed one obvious activity that's missing from GERM's job description: treating patients. That's by design. GERM wouldn't need to replace the rapid response clinical experts like those at Médecins Sans Frontières. Its job would be to coordinate their efforts and complement their work by doing disease surveillance, computer modeling, and other functions. No one from GERM would be responsible for caring for patients.

I put the cost of running the GERM team in the neighborhood of $1 billion a year to cover salaries for 3,000 people plus equipment, travel, and other expenses. To put that number in perspective: $1 billion a year is less than one one-thousandth of the world's annual spending on defense. Given that it would be an insurance policy against a tragedy that costs the world trillions of dollars—as COVID has—and also help drive down the human and financial burden caused by other diseases, a billion dollars a year would be a bargain.* Don't think of this spending as charity or even traditional

* This organization should not be paid for by private citizens. It needs to be accountable to the public and have authority from the WHO.

development assistance. Just like defense spending, it would be part of every nation's responsibility to ensure the safety and security of its citizens.

The GERM team is essential to running a proper monitor-and-respond system, and I will return to it repeatedly in the coming chapters. You'll see the crucial role it will play in every aspect of pandemic prevention: disease surveillance, coordinating the immediate response, advising on the research agenda, and running tests of systems to find their weak spots. Let's turn first to the problem of how you detect an outbreak to begin with.

CHAPTER 3

GET BETTER AT DETECTING OUTBREAKS EARLY

How many times have you been sick in your life? Most people have probably suffered through a number of colds and stomach bugs, and if you're unlucky, you may have caught something worse like the flu, measles, or COVID. Depending on where you live in the world, you may have dealt with malaria or cholera.

People get sick all the time, but not every illness leads to an outbreak.

The task of watching for the cases that are merely troublesome, the ones that could be catastrophic, and everything in between—and ringing the alarm bell when necessary—is known as epidemic disease surveillance. The people who do disease surveillance aren't looking for a needle in a haystack; they're looking for the sharpest, deadliest needles in a mountain of somewhat duller ones.

The term *surveillance* has an unfortunate Orwellian connotation, but in this sense it just refers to the networks of people around the world who keep track of what's happening with health day to day. The information they provide does everything from shaping public policy to informing the decision about which strain of flu you'll be vaccinated against each year. And as COVID has made clear, the world woefully underinvests in disease surveillance. Without a stronger system, we won't detect potential pandemics soon enough to prevent them.

Fortunately, this is a solvable problem, and in the rest of this chapter I'll explain how we can solve it. I'll start with the local health care workers, epidemiologists, and public health officials who are the first people to see evidence of a pandemic in the making. Next, I'll explain some of the obstacles that make disease surveillance difficult for everyone—the fact that many births and deaths are never officially recorded, for example—and tell you about how some countries are overcoming these hurdles.

Finally, I'll explore the cutting edge of disease surveillance: the new tests that will radically change the way doctors detect diseases in their patients, and a novel citywide approach to studying the flu that was pioneered in my hometown of Seattle. (The twists and turns and ethical dilemmas in that story are intense.) By the end of this chapter, I hope to persuade you that, with the right investments in people and technology, the world can get ready to see the next pandemic coming before it's too late.

January 30, 2020, marked a major milestone in the COVID pandemic: The director-general of the WHO declared the disease a "Public Health Emergency of International Concern." That's an official designation under international law, and when the WHO invokes it, every country in the world is supposed to respond by taking various steps.*

Although a few diseases, such as smallpox and new types of flu, are so alarming that they're supposed to be reported as soon as they're detected, most of the time the system operates as it did with COVID. The WHO—trying to protect the public without

* There isn't yet any mechanism for making sure they actually take these steps, though.

causing panic—waits until it has enough data before setting a major international response into motion.

One source of information, as you might expect, is the everyday operations of a health care system: doctors and nurses interacting with their patients. With a few exceptions such as the ones I mentioned earlier, a single case of some disease isn't going to set off alarm bells; most staffers at a clinic won't be unnerved by one individual who shows up with a cough and a fever. Generally, it's the suspicious-looking clusters of cases that will draw attention.

This approach is called passive disease surveillance, and here's how it works. The staff at a clinic pass information about the cases of reportable diseases they're seeing up the chain to their public health agency. They won't share details about each case, but they will give the aggregate numbers of reportable illnesses. From there, ideally, the data will be fed into a regional or global database, which makes it easier for analysts to see patterns and respond accordingly. Countries in Africa, for example, enter aggregated data on certain diseases into something called the Integrated Disease Surveillance and Response system.

Suppose their aggregate data shows an unusual number of pneumonia cases in health care workers. That's a red flag, and hopefully an analyst for a state or national health agency who's monitoring the database will note the spike in cases and mark it for further investigation. In the world's most advanced health systems, the spike might be flagged by a computer system, which then notifies people at the health agency that they need to take a closer look.

Once you suspect there is an outbreak, you need to find out far more than the number of cases. You first need to confirm that the numbers are higher than expected, which requires knowing the size of the population you're dealing with, based on tracking the numbers of births and deaths—a subject I'll return to later in this chapter. If you determine that the disease might spread quickly, you need information such as exactly who was infected, the locations where the infected people might have picked up the pathogen, and the people to whom they may have passed it. Gathering this information can be a time-consuming task, but it's an essential step in disease surveillance, and one of the many reasons that health systems need to be well funded and staffed.

Clinics and hospitals are primary sources of information about the diseases that are passing through a community, but they aren't the only ones. After all, they see only a small fraction of what's going on. Some people who are infected don't feel sick enough to bother going to the doctor, particularly if getting to the clinic is expensive or especially onerous. Others don't have any reason to see a doctor, because they don't feel the least bit sick. And some diseases spread so quickly that it's a bad bet to wait for infected people to show up at the clinic. By the time you notice a jump in cases, it may be too late to stop a big outbreak.

That's why, in addition to monitoring the people who come to clinics and hospitals, it's important to go looking for known diseases by meeting potential patients where they are. That's called active disease surveillance, and a great example is the outreach done by workers in polio campaigns. They go on rounds in the community not only to vaccinate children but also to be alert for kids with polio symptoms, such as unusually weak leg muscles or leg paralysis that can't otherwise be explained. And polio surveillance teams can often do double duty, as they did during the Ebola epidemic in West Africa in 2014–15, when they were trained to watch for telltale signs of Ebola as well as polio.

Some countries are developing smart ways to get even more eyes watching for signs of danger, whether they're coming from a known disease or a new one. Most of the major outbreaks in recent years also showed up in blog posts and social media. This data can be subjective, and there's a lot of noise surrounding the signal, particularly online, but it's often a useful supplement to the insights that health officials get from more traditional indicators.

In Japan, postal workers perform some health services and disease surveillance. In Vietnam, teachers are trained to file a report with local health authorities if they ever notice that several children are absent from school with similar symptoms in the same week, and pharmacists are instructed to raise the alarm when they see a spike in sales of medicines for fever, cough, or diarrhea.

Another relatively new approach is to go looking for signals in the environment. Many pathogens, including poliovirus and coronaviruses, show up in human feces, so you can detect them in the sewage system. Workers take samples of wastewater from treatment plants or open sewers and bring them to a lab, where they're checked for these viruses.

If the sewage samples come back positive, someone will visit the community they came from to identify people who might be infected, step up vaccination efforts, and educate everyone on what to watch for. The idea of checking wastewater was first developed for polio surveillance, but in some countries the technique is also employed to study the use of illicit drugs and the spread of COVID. Studies have shown it can even be part of an early warning system, letting officials prepare for a surge in cases before they show up in clinical test results.

In most rich countries, it is hard to be born or die without the government recording it—the odds are high that the event will be entered into a birth or death registry. But in many low- and middle-income countries, that's not the case.

Many of them estimate the number of births and deaths using household surveys that are held several years apart, which means they don't have precise data—just a wide range of possible numbers. And it may take years for someone's birth or death to be counted in the government's records, if it ever gets counted at all. According to the WHO, only 44 percent of children born in Africa show up in their government's registry. (In Europe and America, more than 90 percent do.) In low-income countries, just one out of every ten deaths is recorded by the government, and only a tiny fraction of those records includes a cause of death. Many communities where births and deaths aren't recorded are essentially invisible to their country's health system.

Given the challenge of recording major life events, it is hardly surprising that many cases of illness in these communities go undetected too. At the end of October 2021, estimates showed that about 15 percent of COVID infections worldwide were being detected. In Europe the rate was 37 percent, but in Africa it was only one percent. With so little precision, and with samples taken only every few years, the death statistics won't help us detect or control an epidemic.

When I first got involved in global health, around 10 million children under the age of five were dying every year, the vast majority of them in low- and middle-income countries. That number was shocking on its own, but even worse, the world knew little about why these children died. Official reports would show a huge percentage of deaths that were simply labeled "diarrhea," but lots of pathogens and conditions can cause diarrhea, and since nobody knew for certain which ones were the main causes of child mortality, we didn't know how to prevent those deaths. Over time, the Gates Foundation and other organizations funded studies that pointed to rotavirus as a major cause, and researchers were able to develop an affordable rotavirus vaccine that prevented more than 200,000 deaths in the past decade and will prevent more than half a million by 2030.

Yet identifying rotavirus as a chief culprit solved only one of the mysteries of child mortality. The places that experience the highest rates of child mortality are also, not coincidentally, the ones that are least well equipped with diagnostics and other tools that might help them understand what happened. A large share of the deaths happen at home, not in a hospital, where the staff could have recorded the child's symptoms. It has taken dozens of studies to gain an understanding of questions like why children die in their first thirty days of life and which respiratory diseases cause the most child mortality.

Mozambique is a good example of how the system can work better. Until fairly recently, the government there counted deaths by surveying small samples of the country every few years and then using the data to estimate nationwide mortality. In 2018, though, Mozambique began building what's known as a "sample registration system," which involves continuous surveillance in areas that are representative of the country as a whole. Data from these samples is fed into statistical models that make high-quality estimates about what's going on throughout the nation. For the first time, Mozambique's leaders can see accurate monthly reports on how many people died, how and where they died, and how old they were.

Mozambique is also one of several countries that are deepening their understanding of child mortality by participating in a program called Child Health and Mortality Prevention Surveillance, or CHAMPS, a global network of public health agencies and other organizations. The genesis of CHAMPS dates back nearly two decades to some of my earliest meetings on global health, when I was hearing from experts about the gaps in the field's understanding of why children die. I remember asking, "What do the autopsies reveal?" and being educated on how impractical they are in developing countries. A full autopsy is an expensive, time-consuming task, and the child's family often won't consent to such an invasive procedure.

In 2013, we funded researchers at the Barcelona Institute for Global Health to refine a procedure called minimally invasive autopsy, or tissue sampling, which involves getting small samples from the child's body for testing. Sometimes, of course, the family members find it too painful to allow a stranger to study their baby in this way. But many agree to the request.

As the name implies, the process is far less invasive than a full autopsy, yet studies have shown that it produces comparable results. Although it is used in only a small number of cases and wasn't created with pandemic prevention in mind—the purpose was to give broader insights into child mortality—the information gleaned from minimally invasive autopsies can give researchers early evidence of an outbreak that is killing children.

I witnessed one of these autopsies during a trip to South Africa in 2016. I had read about how the procedure worked, but I knew that watching one in person would help me understand it in a way that no memo or briefing document ever could. It's an experience that I will never forget.

On July 12, 2016, a baby boy had been born to a family in Soweto, outside Johannesburg. Three days later, he died. His heartbroken parents, hoping to spare other families the same grief, decided to allow doctors to perform the minimally invasive tissue sampling. They also graciously agreed to let me be present when it was done. (I was not there when the request was made.)

At a mortuary in Soweto, I watched as a doctor carefully used a long, narrow needle to remove small samples of tissue from the

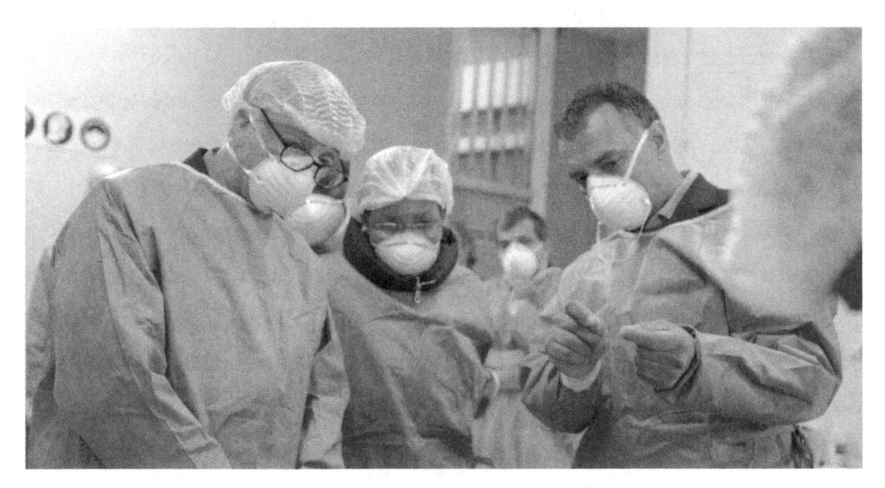

Observing a minimally invasive autopsy in Soweto is a moving experience that I will never forget.

baby's liver and lungs. He also drew a small amount of the child's blood. The samples were stored safely and would later be tested for viruses, bacteria, parasites, and fungal pathogens, including HIV, TB, and malaria. It was over in just a few minutes. Throughout the entire procedure, the medical team treated the boy's body with great respect and care.

The parents were informed confidentially of the results. I never met them, but I hope they got some answers about what happened to their son, as well as a small measure of solace from the fact that their decision to participate in CHAMPS contributed meaningfully to the world's efforts to save children like him.

Today, data on more than 8,900 cases from the CHAMPS network is giving researchers valuable insights into child mortality. The minimally invasive autopsy, and the systemic improvements that Mozambique and other countries are making, are deepening our understanding of why people die. We need to expand these innovative approaches to understand even better how we can intervene to save lives.

Most people will never take a monthly household survey about births and deaths or have anything to do with a CHAMPS-type network. But during COVID, and during major outbreaks in the future, we want to sample the community to find out how many asymptomatic or unreported cases of sickness are out there. The field of diagnostics is rife with innovation to make the process cheaper and simpler—and therefore easier to implement at the scale that will be necessary—so let's go through the state of play and see what's in the offing. I have to make some broad generalizations, because the utility of different tests depends on, among other things, the pathogen you're looking for and the route it takes to get into your body.

Since the beginning of COVID, the U.S. government alone has approved more than 400 tests and kits for collecting samples. Early in the epidemic, you may have become familiar with PCR tests, most of which required the brain-tickling swab way up in your nose. If you've got a COVID infection, the virus will be in your nostrils and in your saliva, and the swab will catch a sample of it. To analyze your swab, a lab technician will mix your sample with substances that make extra copies of any genetic material from the virus. This step ensures that if there's even a small amount of virus in the sample, it won't escape detection. (It is this process of duplication, which mimics the way nature copies DNA, that gives the polymerase chain reaction its name.) A dye is also added, and if the viral genes are present, the dye will begin to glow. No glow, no virus.

Creating a PCR test for a new pathogen is a pretty easy task once you've sequenced its genome. Because you already know what its genes look like, you can create the special substances, dye, and other necessary products very quickly—which is why researchers were able to establish PCR tests for COVID just twelve days after the first genome sequences were published.

Unless the sample is contaminated, a PCR test is unlikely to give you a false positive—if the result says you're infected, you almost certainly are—but sometimes it can return a false negative, meaning that it says you're free and clear even though you're not. This is why if you're experiencing symptoms and get a negative PCR test, you may be asked to take the test again. The test may also pick up on genetic bits of the virus that remain in your blood or nose long after you've been sick, so you may test PCR-positive even if you're not infectious anymore.

The main downside of PCR tests, though, is that they have to be run on special machinery in a lab, which makes them impractical in many parts of the world. The analysis itself takes only a few hours, but if there's a backlog—as there often has been during COVID—you may not get your results for days or even weeks. Given how easily COVID is passed from one person to another, any test result you get more than forty-eight hours after giving the sample is useless: If you're going to spread the virus, you've already done it by then, and if you need to start treatment with an antiviral or antibody medication, you have to do it within just a few days of the infection.

The other main category of tests looks not for the virus's genes, as PCR machines do, but for specific proteins on its surface. These proteins are known as antigens, and so the tests are called antigen tests. They're less accurate, but not egregiously so; they're especially good at detecting when you're capable of infecting others, and they produce results in less than an hour (and often within fifteen minutes).

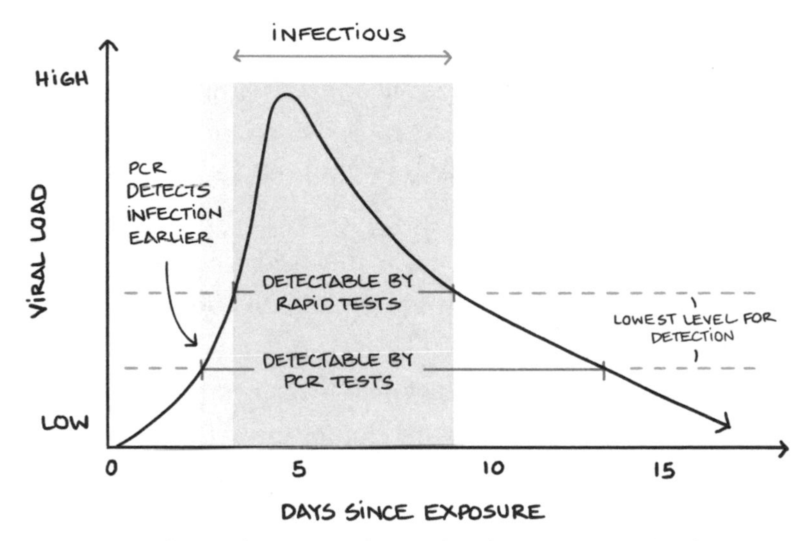

PCR tests detect the virus earlier and pick up lower levels of the virus than rapid (antigen) tests. But they can also return a positive result long after you've stopped being infectious.

Another benefit is that most antigen tests can be taken by anyone in the privacy of the home. If you've ever taken a pregnancy test by peeing on a stick and watching for the plus or minus sign to appear, you've used a thirty-year-old technology called a lateral flow immunoassay—named that way, I suppose, because "test that uses liquid flowing over a surface" was too easy to understand. Many antigen tests work the same way.

During an outbreak, we've got to make it easier for everyone to get tested and get results fast—especially if it's a disease that you can transmit to others before you have any symptoms. And by "we," I mean the United States first and foremost. Other countries, including South Korea, Vietnam, Australia, and New Zealand, far outpaced the U.S. in testing and returning results, much to their benefit.

Ideally, in the future, everyone's results will be linked to a digital data system—with the proper safeguards for privacy—so that public health officials can see what's going on in their community. It's especially important to identify the people who are most likely to spread the infection, since studies have shown that some COVID patients pass the virus to many others, while many COVID patients don't even infect people they're in constant contact with.

Ultimately, we need diagnostic tools that are accurate, accessible for many people around the world, and quick to produce results that feed into the public health system. So let me tell you about some of the exciting work that's going on in this field, with my usual bias for innovations that will benefit people in poor countries as well as wealthier ones.

The one I'm most enthusiastic about comes from the British company LumiraDx, which is developing machines that test for multiple diseases and are so easy to operate that they don't have to be limited to laboratories—they can be used in pharmacies, schools, and other settings. Like antigen tests, they provide rapid results, but unlike antigen tests, they hit the mark about as well as PCR machines do, and at about a tenth of the cost. A single production line can manufacture tens of millions of tests in a year, and a new test for an emerging pathogen can be developed with little or no retooling.

In 2021, a group of partners that included a nonprofit called the African Medical Supplies Platform provided 5,000 LumiraDx machines to countries throughout Africa. That's a tiny fraction of the need, though, and I hope more funders step up.

For now, PCR tests remain the gold standard in terms of accuracy, but they're also slower and more expensive than other methods. Several companies, though, are out to change this through a process called ultra-high-throughput processing—basically, using robotic machines to exponentially increase the number of PCR tests that can be processed in a given time and with a fraction of the workforce.

The fastest one I'm aware of is called Nexar, developed by Douglas Scientific more than a decade ago, but not for anything related to diagnosing diseases in humans—it was originally intended to identify genetic changes in plants that would make them more beneficial as crops. The machine puts hundreds of samples and reagents on a long piece of tape—picture something like a filmstrip—and seals it. The tape goes into a water bath and then, after a couple of hours, is run through a second machine, which scans all the samples and flags the ones that are positive. Like LumiraDx, this system is flexible enough to be able to add new tests quickly and can even use one sample to test for many different pathogens at the same time. For instance, you can test one nasal swab for COVID, flu, and RSV (respiratory syncytial virus) simultaneously, all for a fraction of the cost of current tests.

Amazingly, the Nexar system can process 150,000 tests per day, which is more than ten times what the largest high-throughput processors can do today. The company LGC, Biosearch, which now makes the Nexar machine, is planning several pilot projects to see how it can work with samples collected from various places, including prisons, elementary schools, and international airports. Other

The Nexar™ machine from LGC, Biosearch Technologies.

companies are working on different approaches, and I hope they all keep competing to make cheaper, faster, and more accurate tests. This is an area where the world still needs a lot of innovation.

In short, we need to be able to design a new test very quickly that can be used in many different settings, including clinics, homes, and workplaces—and then, once that test is designed, we need to be able to make many millions of them at ultra-low cost (perhaps less than a dollar per test).

The Seattle area, where I live, has become something of a hub for the study of infectious diseases. The University of Washington has an excellent global health department and one of the best medical schools in the country. The university is home to the Institute for Health Metrics and Evaluation, which I mentioned in Chapter 1. The Fred Hutchinson Cancer Research Center, though focused primarily on cancer, also has top experts in infectious diseases. (It's high-profile enough that it's known around town as Fred Hutch, or simply the Hutch.) PATH is a leading nonprofit dedicated to making sure that innovations in health reach the world's poorest people.

Put so many intelligent people who are passionate about the same field into one city, and they are practically guaranteed to start kicking around ideas. Over the past few decades, Seattle has become home to a thriving, informal network of researchers who exchange ideas within and across institutions.

It was through this network that, over the summer of 2018, a handful of people in genomics and infectious diseases came to a joint realization. Although they represented different institutions—Fred Hutch, the Gates Foundation, and another group called the Institute for Disease Modeling*—they were all worrying about the

* The Institute for Disease Modeling is now part of the Gates Foundation.

same problem: outbreaks of respiratory viruses. These outbreaks kill hundreds of thousands of people every year and are the most likely candidates for a pandemic, but the field needed to learn a lot more about how they move through communities. And the tools at scientists' disposal were, at best, limited.

For example, researchers have access to case counts from hospitals and clinics, but those statistics represent only a small share of the total. The Seattle scientists talked about how they needed to know much more before they could understand how a virus like influenza spreads through a city—most important, they needed to know how many people actually got sick, not just how many were tested. And in an emergency outbreak, city officials would need to quickly identify the lion's share of people who might be sick, get them tested, and inform them about their results. But there was no systematic way to do any of those things.

Eventually, in June 2018, a few of the people who were driving this conversation met with me at my office outside Seattle to explain the problem as they saw it. They outlined a three-year project they called the Seattle Flu Study—a prototype for a citywide effort that could transform the way respiratory viruses were detected, monitored, and controlled—and asked whether I would fund it.

Here's how it would work. Beginning that fall, as the flu season ramped up, volunteers throughout the Seattle area would be asked to answer a few questions about their health. If they'd had at least two symptoms of a respiratory problem in the past seven days, they would be asked to submit a sample that would be tested for a range of respiratory diseases. (Despite the name of the project, it wouldn't be limited to flu—the tests would actually cover twenty-six different respiratory pathogens.)

Some people would give samples at kiosks set up in Sea-Tac Airport, the University of Washington campus, homeless shelters, and a few workplaces around town, but most of the samples would come from local hospitals that had already collected them from patients for other reasons. This is a common practice in medical research: When you get tested at a hospital, the results help the doctor decide how to treat you, but the mucus from your nasal swab might be saved for later. Researchers can then use that sample, after certain private data about you has been removed, to test for other pathogens and understand what's happening throughout the community. Just by being sick, you're contributing to science.

In the Seattle Flu Study, the idea was that all the samples gathered from hospitals and public places would be tested. When one tested positive for the flu, the case would be marked on a digital map showing, nearly in real time, where the known flu cases were. Then the virus would go through a further step: Its genetic code would be studied and compared with genes from other flu viruses found around the world.

This genetic work would be a key part of the Seattle Flu Study, because it would let the scientists understand how different cases were related to one another. How do different flu strains enter the city? If there was an outbreak at the university, how far would it spread within the community?

Genetic information is so useful to epidemiologists due to a fortuitous flaw in the way genes work. Every time a pathogen makes a copy of itself (or forces the host cell to do the copying, as a virus does), it duplicates its genetic code, or genome. The genomes of all living things are made up of just four building blocks, which we represent as As, Cs, Gs, and Ts.* If you're a movie fan, you might remember a sci-fi film with Uma Thurman and Ethan Hawke about genetically enhanced humans, the title of which—*Gattaca*—is a clever arrangement of these building blocks.

The genome gets passed down from one generation to the next, ensuring that children resemble their biological parents. It is what

* RNA viruses actually have Us instead of Ts, but the two substances are functionally identical, so I am sticking with Ts for the sake of simplicity.

makes a person a person, a virus a virus, and a pomegranate a pomegranate. COVID's consists of about 30,000 As, Cs, Gs, and Ts, while yours and mine consist of several billion, but complex organisms don't necessarily have larger genomes. Most ingredients in an ordinary salad have a bigger genome than humans do.

The process of copying genes is imperfect, and it always introduces a few random mistakes, especially in viruses like COVID, influenza, and Ebola. Some As get copied over as Cs, and so on. Most of these mutations either have no effect or leave the copy unable to function, but once in a while they make the copy better suited to survive in its environment than the original that produced it. This is the evolutionary process that leads to COVID variants.

Figuring out the order in which an organism's genetic letters appear is what's known as *sequencing its genome.* By sequencing the genomes of many different versions of a virus and studying the different mutations among them, scientists can construct what amounts to its family tree. At the bottom of the tree is the latest generation. Further up the tree are that generation's ancestors, all the way up to the first known specimen. The places where the tree branches split indicate major evolutionary steps, such as the emergence of a new variant, and the tree can even be used to record related pathogens that have been found in animals and might make the leap to humans.

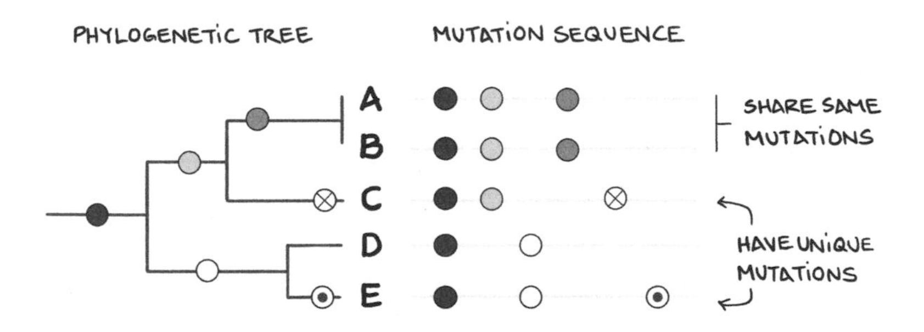

All this family tree information, in combination with a good testing regime, can provide invaluable insights into how a disease moves through a community. In South Africa, for instance, a good testing system paired with genetic analysis of HIV revealed that many young women who were living with the virus had acquired it by having sex with older men—information that led to changes in the way the country approached HIV prevention. More recently, genetic sequencing revealed that a 2021 outbreak of Ebola in Guinea started with a nurse who had been infected, astonishingly, five years earlier. Scientists were stunned to learn that the virus could remain dormant for so long, and based on this new information, many are now rethinking ways to prevent Ebola outbreaks.

The problem that the Seattle scientists and their colleagues kept running up against was that in the United States key parts of the infrastructure needed for this kind of analysis were missing.

Think about the way we handle influenza here. Most people who think they have the flu don't bother going to the doctor—they just stock up on over-the-counter drugs and sweat it out. If they do end up at a clinic, the doctor might make a diagnosis based on the symptoms alone, without running a test. The cases that get reported to public health officials are the ones in which a test is ordered by a doctor who works at a clinic that is participating in a voluntary influenza-reporting program.

The fact that so few tests are run has a ripple effect: Too few samples of the influenza virus are sequenced. Plus, many of the ones that do get sequenced are not accompanied by information about the people they came from—where they live, how old they are, and so on. You could have a million sequences of a virus, but if you don't know anything about the people who provided them, you won't be able to figure out where the disease started or how it spread from one place to another.

The Seattle Flu Study was designed to tackle this problem head-on. Not only would it create a system for testing lots of volunteers and sequencing lots of viral genomes, but—subject to privacy safeguards—the sequencing data would be linked to information about the people it came from. And the near-real-time, citywide flu map that the project was going to create would be a game changer for detecting and stopping outbreaks.

I thought the Seattle Flu Study was an ambitious and unique idea, and it had a chance to make progress on some of the problems I had called out in my TED talk years before. I agreed to fund it through the Brotman Baty Institute, a research partnership between Fred Hutch, the University of Washington, and Seattle Children's.

The team quickly got to work on the infrastructure they had envisioned. They created a system to develop and prove a new diagnostic test, process and share the results, and perform quality checks to make sure all the work was valid. In the second year, they added a way to let participants take their own samples at home and return them in the mail. With this innovation, the Seattle Flu Study became the first medical study anywhere with a soup-to-nuts process that let people order a kit online, have it delivered to their home, ship it back to the lab, and receive a result. This was pioneering work and a point of pride for the team, but none of us had any idea just how crucial it was about to become.

In 2018 and 2019, the Seattle Flu Study tested more than 11,000 cases of flu and sequenced more than 2,300 influenza genomes—about one sixth of all flu genomes sequenced anywhere in the world in that time. They were able to show that flu in Seattle wasn't one homogenous outbreak, but actually a series of overlapping outbreaks of different flu strains.

Then, in the first few weeks of 2020, everything changed. Nearly overnight, influenza was no longer the virus we needed to worry about the most. The scientists who had spent countless hours planning and creating the flu study were now thinking about nothing but COVID.

By February, a genomics researcher named Lea Starita had developed her own PCR test for COVID, and her team began running it on a few hundred samples that they had gathered for the flu study.

Within two days, they had found a positive case, a sample submitted to the study by a local clinic that had treated a patient for flulike symptoms.

After sequencing the virus from this positive sample, one of the team members—a computational biologist named Trevor Bedford—made a disturbing discovery: Genetically, it was closely related to another, earlier case in Washington state. After comparing mutations in the genomes of the two viruses, Bedford inferred that the viruses were closely related.* It was proof of what many scientists had suspected—that COVID had been moving throughout the state for quite some time.

Then the group turned to the next logical question: Based on what they knew about the two cases they had sequenced, and how long they now knew the virus had been circulating, how many more people might be infected? A disease modeler named Michael Famulare ran the calculations and put the estimate at 570.†

At the time, only eighteen COVID cases had been confirmed through testing in all of Western Washington. With their work, Bedford, Famulare, and their colleagues showed that the country's COVID testing system was utterly inadequate. In Washington state alone, hundreds of people had COVID but didn't know it, and the disease was spreading fast.

But there was a catch: They weren't sure they could tell anyone what they knew.

The patient at the clinic who had given the sample didn't know that it had been used in a research trial. Although it was standard

* Subsequent evidence has muddied the waters as researchers have sequenced other samples from around that time. We may never know for sure whether the virus in the second case descended from the one in the first case. But there is universal agreement that the researchers made the right inference given the information available and that there was a lot of transmission happening at the time.

† To be more precise, Famulare put the number at 570, with 90 percent confidence that it was between 80 and 1,500.

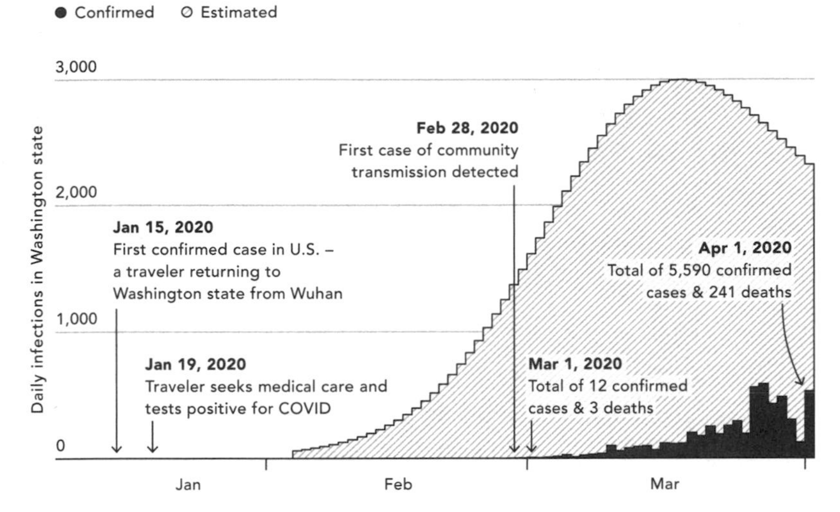

When COVID arrived in Washington state. Seattle Flu Study scientists found that hundreds of people likely had undetected cases of COVID. This chart shows the difference between confirmed cases of COVID and the estimated number of people infected in the first three months of 2020. (IHME)

practice to test the patient's sample for another disease such as COVID, revealing the results of that test to anyone—even to the patient, much less to public health officials—was another matter. It would be a violation of the flu study's research protocols.

Also, their COVID test had been approved for use in research studies, but not medical settings, in which the results are given to patients. Although the study team had been talking to government regulators for weeks, there was no way to get their test authorized for medical use. The rules for okaying COVID tests developed by anyone other than the CDC had not even been written yet.

It was a difficult dilemma. On the one hand, revealing the results would violate the standards under which they were operating as ethical researchers and could run afoul of government rules.

On the other hand, how could the team withhold test results from someone carrying a virus that was causing a pandemic? Or from public health officials who needed to know that COVID was spreading in the state and had almost certainly infected hundreds more people than they realized?

One member of the group clarified the debate with a simple question: "What would a reasonable person do?" When he put it like that, the answer seemed obvious. A reasonable person would protect the individual and the community by revealing the results. So they did.

The news made a splash. "Coronavirus May Have Spread in U.S. for Weeks, Gene Sequencing Suggests," as *The New York Times* put it.

Although the decision generated some blowback from government regulators and the team had to temporarily stop testing the hospital samples, I felt (and still feel) that they did the right thing. The University of Washington review board that was overseeing the project came to the same conclusion, noting that the team's actions were the responsible and ethical thing to do. And state and federal officials continued to work with them on ways to study COVID in the area.

In March 2020, the flu study group teamed up with the public health agency in King County, where Seattle is located, to create the Seattle Coronavirus Assessment Network, or SCAN. The pioneering system they had set up for gathering and processing flu samples and informing people of their results would be put to a new use: testing as many people for COVID as they could, mapping the results, and adding to the world's collection of genetic sequencing for this brand-new pathogen.

SCAN's efforts got a huge lift from another group of local researchers, who showed government regulators that twirling a swab in the tip of a person's nose produced results that were just as good as the brain-tickling jab that other COVID tests required. That was a major advance, because it let people swab themselves, while the previous approach had to be handled by a health care worker. It was also much less uncomfortable, removing a barrier that had kept some people from getting tested. The old way inevitably made the test subject cough, which raised the odds of exposure for whoever was administering it, and the world was in the unprecedented situation of running out of the longer swabs.*

From March to May, things ran as smoothly as could be expected during a pandemic. The SCAN team collected samples from volunteers, told them whether they had COVID, started building a map of cases, and made sure the positive samples were sequenced. During that time, SCAN accounted for as much as a quarter of all testing in King County, and its maps helped local officials understand where the disease was most prevalent.

Then, in May, the federal government suddenly ordered them to stop. The team had run up against another problem: whether they were allowed to test samples that people had collected on their own (rather than having a health care worker do it). Up to that point, the federal government's rules regarding who could test self-collected samples had been murky. When they did clarify the rules, it was bad news for SCAN: They needed federal approval for their test. The team immediately started scrambling to find another way forward.

Then, two weeks later, the FDA changed its policy again. Researchers could test samples gathered by participants as long as they had approval from the review board overseeing the work. SCAN got the okay from theirs, and on June 10 the program took up testing again.

Over the rest of that year, the team toted up several accomplishments. They processed almost 46,000 COVID tests, nearly all of them from people who had enrolled online from home (as opposed to the kiosks in public places, which had largely shut down). They sequenced nearly 4,000 COVID genomes, more than half of all the sequences in Washington state that year, and they advised teams

* It took a long time for the new approach to get picked up. As I write this, relatives are still asking me: "How come they're jamming the swab into my brain? I thought you said they got rid of that?" The reason is that each time a test is approved by government regulators, the swab has to be approved too—even if it has been successfully used in other tests.

who were setting up similar studies in Boston and the San Francisco Bay Area.

As I write this toward the end of 2021, SCAN is still running, and the Seattle Flu Study continues to gather data on influenza and two dozen other pathogens. Trevor Bedford, the researcher who discovered the genetic similarities between the two COVID samples and figured out their importance, is widely recognized for his groundbreaking contributions to COVID science. His genomic family trees are used throughout the world, and he has become an excellent public communicator, breaking down complex matters of epidemiology and genome sciences for his hundreds of thousands of Twitter followers.

The U.S.—and, really, any country with a similarly patchwork system of testing and sequencing—needs to invest in many more projects that build on what the Seattle team has learned. One lesson is that we need to be setting up response systems well ahead of the next big outbreak, as the Seattle Flu Study and SCAN tried to do. Governments need to establish working relationships with infectious-disease experts in the public and private sectors. Regulations need to allow for the rapid approval of tests when a pathogen emerges that we've never seen before. America's world-class research institutions and its private diagnostics companies have incredible talent and capacity to help, but they should be able to get involved right away, without jumping through all the hoops that the SCAN team went through.

The countries that get this done will be well positioned in the next major outbreak. It's not a coincidence that South Africa, a country that has spent decades investing in testing and sequencing for its fight against HIV and TB, was the first to identify at least two major COVID variants.

There are some innovations coming in genomic sequencing equipment that will help a lot. For example, Oxford Nanopore, a spinoff from Oxford University, has developed a portable gene sequencer that eliminates the need for a full laboratory. It does require an online computer with a powerful processor, but some researchers from Australia and Sri Lanka are working on solving that problem too: They created an app that allows the information from the sequencer to be processed offline, on a standard-issue smartphone. In one test, the app/sequencer combination was able to sequence COVID genomes from two patients in less than thirty minutes each. Oxford Nanopore is now working with the Africa CDC and other partners to deploy similar advances throughout the continent.

Another lesson is that setting up a platform similar to SCAN or the Seattle Flu Study—that is, making the test, creating the website where people can sign up, processing their samples, etc.—is only part of the challenge. It is another matter altogether to make sure that the results reflect the actual makeup of the community. Not everyone can navigate a website easily. Language barriers can get in the way. When demand for test kits is high and supplies are limited, people who can stay at home repeatedly checking a website have an advantage over essential employees who still have to go to work. Bridging these gaps has been a challenge in Seattle, and anyone who's looking to do something similar should keep them in mind. Making the most of technical advances takes a strong public health system that is trusted by people throughout the community.

On a list of jobs that are both super-important and super-obscure, I'd probably put "disease modeler" near the top. Or at least I would have before 2020. Once COVID came along, people who had been toiling in obscurity for decades found themselves thrust into the limelight. Disease modelers traffic in predictions, and during a pandemic there are few things that news reporters love more than a prediction.

Most of my experience with disease modeling comes from my

work with IHME and with the Institute for Disease Modeling, or IDM, the group that was involved in the Seattle Flu Study. But there actually are hundreds more models being run by researchers all over the world, and different ones can help answer different types of questions. I'll give you two examples.

One is the work on the Omicron variant that was done in late 2021 by the team at the South African Centre for Epidemiological Modelling and Analysis (based in Stellenbosch, South Africa). At the time, researchers had identified Omicron, but they hadn't yet answered some crucial questions about it, including: "How often is Omicron reinfecting people who have already had an earlier variant of COVID?" Using a database that tracks cases of infectious diseases from throughout the country, the South African team found the answer: Omicron was far more capable of reinfecting people than previous variants were. This and other work by the team showed that, unlike other variants that had fizzled out, Omicron was likely to spread fast anywhere it landed—which is exactly what happened.

Other modeling teams tackled different questions. A group based at the London School of Hygiene & Tropical Medicine, for example, quantified the impact of masks, social distancing, and other methods of slowing transmission. And in 2020, their models produced some of the most accurate and timely forecasts of how the virus would spread in low- and middle-income countries. (In fact, they often outperformed IDM, the group that's now part of the Gates Foundation—and the IDM team would be the first people to tell you that.)

To get an idea of what modelers do when they're trying to predict pandemic patterns, think about forecasting the weather. Meteorologists have models that are pretty good at predicting whether it will rain tonight or tomorrow morning. (If it's winter in Seattle, the answer is bound to be "yes.") Their models are less accurate for ten days from now, and they have no idea what precisely will happen six or nine months from now.* Modeling diseases with variants is a little like that, and though it will never be a perfect science, it will eventually do better than the weather forecast.†

What a modeler tries to do, in essence, is to analyze all the available data—the sources I've described in this chapter, and many others, such as mobile phone data and Google searches—for two purposes. One is to determine why something happened in the past, and the other is to make an educated guess about what might happen in the future. It was computer modeling that showed, early on, that even if only 0.2 percent of the population got infected with COVID, hospitals would be overflowing with patients in no time.

Disease modeling also has broader benefits for people who study public health. It forces them to lay out all of their assumptions and data, which highlights what they know, what they don't know, and how certain they are. It also lets them study which characteristics of the disease and our response might have the biggest impact in the future: For example, what are the benefits of vaccinating high-risk people before the rest of the population? If a variant emerges that's ten times more transmissible, how many more cases, hospitalizations, and deaths should we expect? How much would it help if a certain percentage of people mask up?

For me, one of COVID's chief lessons about modeling is the extent to which every model relies on good data, and just how hard it can be to get that data. How many tests are being done? How many are positive? Disease modelers had all sorts of trouble finding out. Some U.S. states didn't break down their cases by location

* Although it is certain that global temperatures are going up, which will have terrible consequences if we don't act.

† IHME was criticized early in the pandemic for making forecasts that were too optimistic and for not emphasizing the uncertainty surrounding their projections. But they are listening to the feedback and improving their work, as good scientific organizations do all the time.

or demographics. Sometimes reporting would pause over a holiday weekend, and then all the cases would get submitted on the first day people were back in the office, and the modelers would be left to estimate what had really happened.

I also couldn't help noticing how often news reports about some modeler's latest findings would leave out important nuances and caveats. In March 2020, Neil Ferguson, a highly respected epidemiologist at Imperial College, predicted that there could be more than 500,000 COVID deaths in the U.K. and more than 2 million in the U.S. over the course of the pandemic. That caused quite a stir in the press, but few reporters mentioned a key point that Ferguson had been very clear about: The scenario of his that made all the headlines assumed that people wouldn't change their behavior—that no one would wear masks or shelter in place, for instance—but of course that wouldn't be the case in reality. He wanted to show how high the stakes were and demonstrate the value of masks and other interventions, not drive everyone into a panic.

The next time you hear about some prediction made by a disease modeler, keep a couple of things in mind. First, every variant is different, and it's hard to predict the severity of each one until there are several weeks of data available. Second, all models have limitations, and the report you're hearing might have left out some important caveats. The level of uncertainty, for example, can be quite high. Remember Mike Famulare's estimate of 570 cases in Washington state, with a 90 percent certainty that it was between 80 and 1,500? Any report that omitted the range of possibilities left out some pretty important context.

Finally, everyone involved in creating disease models should think about how people will use their work, and they should try to communicate clearly so they lower the odds that it will be misunderstood or misused. Disease modeling needs to be done with a healthy dose of modesty, especially if the forecast stretches beyond four weeks or so.

—

What everything in this chapter adds up to, I believe, is a clear agenda for the kind of disease surveillance that we need in order to prevent pandemics.

One step is to invest in all the elements of a robust health system that make it possible to detect and report diseases, as well as treat them. That's especially true in low- and middle-income countries, whose health systems are often underfunded. If doctors and epidemiologists don't have the tools and training they need, or their national health agency is weak or nonexistent, we're going to keep seeing outbreak after outbreak. Every community in every country should be able to detect an outbreak in seven days or less, report it and begin investigation within another day, and implement effective control measures within another week—standards that will give everyone in a health system goals to aim for and ways to measure their improvement.

Another step is to expand on efforts to understand the causes of death in adults and children alike. This work will have a double benefit, giving us new insights into health and illness as well as yet another window into emerging threats.

Third, we need to know the enemy we're up against. So governments and funders should back innovative ways to test mass numbers of people in a short time—especially high-volume, low-cost tests designed to work in low- and middle-income countries. New tests should make it possible to link the results back to the patient, with privacy safeguards, so the data can inform both individual care and public health measures. Genetic sequencing needs to expand dramatically. In addition, we need to continue to study how viruses evolve in animals and learn more about which ones might cross over to humans—after all, out of the thirty most recent unexpected outbreaks, three quarters have involved animals (other than humans). And in a major outbreak, when tests may be in short supply, maps that chart the prevalence of the disease should inform who gets priority—so that tests go to the people with the highest odds of being infected.

Finally, we need to invest in the promise of computer modeling. The analyses produced during COVID have been extremely helpful, but they can be better. More data, more accurate data, and constant feedback on their models will make all of us safer.

CHAPTER 4

HELP PEOPLE PROTECT THEMSELVES RIGHT AWAY

GREETING ANXIETY

I'm confused about what to do when I meet someone these days. Should we bump fists, shake hands, or just smile and wave? Depending on the nature of our relationship, I might want a combined handshake-and-hug, particularly if we haven't seen each other for months.

Of course, navigating hellos and goodbyes is just one of the many ways that COVID complicated our social interactions. Should you stay home if you've been exposed? Who should wear a mask, and when? Is it okay to have a party, can it be indoors or outside, and how far apart do people need to stand? Should you wash your hands more often? What about large public gatherings and public transit—should they keep going? Can schools, offices, and retail businesses stay open?

Although not all of these decisions are up to individuals, many of them are. And during a pandemic, when your options seem more limited than ever, making a choice can be empowering. Even if you're not in a position to help scientists find a cure or a vaccine, you can still choose to wear a mask, stay home if you're feeling sick, and postpone your big parties.

It is unfortunate that in some places, especially in the United States, people have resisted making choices that will keep them and their families safer. I don't agree with these choices, but I also think it's unhelpful to simply label them "anti-science," as so many people do.

In her book *On Immunity,* Eula Biss looks at vaccine hesitancy in a way that I think also helps explain the resentment we're seeing toward other public health measures. The distrust of science is just one factor, she says, and it is compounded by other things that trigger fear and suspicion: pharmaceutical companies, big government, elites, the medical establishment, male authority. For some people, invisible benefits that might materialize in the future are not enough to get them past the worry that someone is trying to pull the wool over their eyes. The problem is even worse in periods of severe political polarization, such as the one we're in now.

It didn't help that, when COVID first hit, there wasn't enough evidence to weigh the costs and benefits of different measures. It was especially hard to make the case for painful measures like closing businesses and schools. Many of these steps hadn't been widely used since the 1918 pandemic, and while the costs associated with them were predictable and immediately apparent to anyone who thought about it, the precise benefits—especially given that we were dealing with a new pathogen—were not.

Part of the problem is that it's quite difficult to assess the impact of many of these measures—which are broadly called "nonphar-

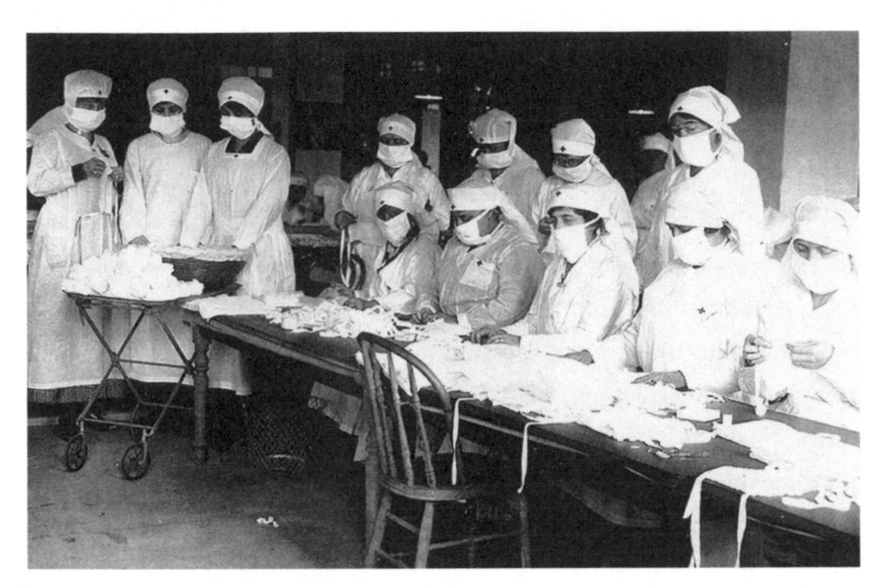

Red Cross volunteers in Boston, Massachusetts, assemble masks made from gauze to prevent the spread of influenza during the 1918 pandemic.

maceutical interventions," or NPIs—in a controlled environment. Drug and vaccine trials are expensive and time-consuming (as I'll explain in later chapters), but they allow us to run experiments that test the effectiveness of drugs and vaccines. By contrast, no one's going to run an experiment that shuts down all the schools and businesses in a city just to measure the costs and benefits.

Now, after two years of studying NPIs in the real world, we know a great deal about their effectiveness, at least for COVID. The pandemic gave us real-world learning that no experiment ever could. At almost every level of government—city, county, state, provincial, and federal—officials have looked at data to see what was working, and thousands of academic studies have documented the impact of various NPIs. These efforts have dramatically improved our understanding of the field. The variations in policies across similar cities or countries allowed researchers to isolate the impact of individual NPIs in ways that had never been possible before.

This is good news, because NPIs are our most important tool in the early days of an outbreak. There's no lab time required to put mask mandates in place (assuming we can provide the masks), figure out when to cancel big public events, or limit how many people can sit in a restaurant. (Though we will need to make sure that whatever NPIs we deploy are appropriate to the pathogen we're trying to stop.)

These interventions are what we use to flatten the curve—that is, to slow down transmission so that hospitals don't get overwhelmed with patients—without having to identify everyone who's infected. If you catch an outbreak soon enough, you can find almost all the people who are infected and test everyone they have come in contact with. That's crucial, especially since it is notoriously difficult to find people who are carrying the pathogen but aren't showing symptoms; NPIs help prevent them from spreading COVID just as well as people who are symptomatic.

I don't mean to suggest that NPIs are a painless solution. While some, such as wearing masks, have few downsides for most people (aside from steamed-up lenses for those of us who wear glasses), others—like closing businesses and stopping big public gatherings—have a huge impact on a society and implementing them is a massive undertaking. But we can do them right away, and now we know how to do them better than we did before.

Let's go through some of the top insights from the past two years.

"If it looks like you're overreacting, you're probably doing the right thing."

That's a quote from Tony Fauci, and I agree. The irony of NPIs is that the better they work, the easier it is to criticize the people who put them in place. If a city or state adopts them early enough, the case numbers will stay low, and critics will find it easy to say they weren't necessary.

For example, in March 2020, officials in the city and county of

St. Louis took several steps to limit transmission, including a shelter-in-place order. As a result, the initial outbreak in St. Louis was not as severe as it was in many other U.S. cities, leading some to suggest that the policies were an overreaction. But one study found that if the government had implemented the very same interventions just two weeks later, the number of deaths would have shot up sevenfold. St. Louis would have been on par with some of the hardest-hit areas in the country.

Nor was that the first time that St. Louis had led the way: Virtually the same thing had happened a century earlier. Shortly after detecting its first flu cases during the 1918 pandemic, the city closed schools, banned large public gatherings, and put social distancing measures in place. Philadelphia, on the other hand, waited to do these things. For two weeks after its first case, it allowed large public gatherings, including a citywide parade.

As a result, Philadelphia's peak death rate was more than eight times higher than St. Louis's. Later, studies found that the pattern held across the country: Cities that put multiple measures in place, and did it early, had death rates that were half that of cities that had waited.

Compare countries instead of cities, and you'll get similar results. During the first wave of COVID, Denmark and Norway implemented strict lockdowns early on (when fewer than thirty people in each country had been hospitalized), while the government of neighboring Sweden relied more on recommendations than requirements, keeping restaurants, bars, and gyms open and only encouraging but not requiring physical distancing. One study found that if Sweden's neighbors had followed its lead instead of locking down stringently, Denmark would have had three times as many deaths as it did during the first wave, and Norway nine times as many as it did. Another study estimated that NPIs in six large countries, including the United States, prevented nearly half a billion COVID infections in the first few months of 2020 alone.

Not only should you appear to overreact at first, as Tony Fauci said, but you also have to be careful about relaxing all NPIs too soon. When the most effective public measures are relaxed—such as restrictions on large gatherings—case numbers tend to go back up (all other things being equal). The problem with relaxing these measures early is that you have large numbers of people who are what experts call "immune naïve": They've never been exposed to the virus—and they're susceptible to infection. Just as it's important to keep taking antibiotics for a bacterial disease even when you start feeling better, in some cases we need to keep some NPIs going until we can develop medical tools that protect you from infection and keep you out of the hospital if you do get sick. Or at least until we're able to reduce transmission dramatically by testing lots of people and isolating the positive or suspected cases, as South Korea did.

Also, not all overreactions—or apparent overreactions—are created equal. Closing borders, for example, did slow the spread of COVID in some regions. But border closures are a hammer that needs to be wielded very carefully. By cutting off trade and tourism, they can crater a country's economy so badly that the cure becomes worse than the disease. This is particularly true if, as is often the case, the border controls come too late. And they create a disincentive to report an outbreak early; South Africa, for example, was sanctioned with a travel ban when it identified the Omicron variant, even though some other countries where Omicron was spreading weren't treated the same way.

Even though lockdowns have clear benefits for public health, it's not always clear whether in lower-income countries they are worth the sacrifice. In such places, closing down sectors of the economy can lead to acute hunger, drive people into extreme poverty, and increase deaths from other causes. If you're a young adult and spend your day working outside—as many people in low-income countries do—COVID will not seem as scary as the possibility of not having enough food to feed your family. As I'll explain later in this chapter, there's a similar phenomenon in wealthier countries: Low-income people in those places are both less likely to be able to comply with lockdowns and more likely to be affected by COVID.

With the benefit of hindsight, we know that in a lot of places—at least when COVID was at a peak—the cost of not locking down would likely have been even higher. The economy was bad when businesses shut down, but it could have been even worse if the virus had been allowed to run rampant and kill millions more people than it already had. By saving lives, lockdowns can make it possible to start the economic recovery sooner.

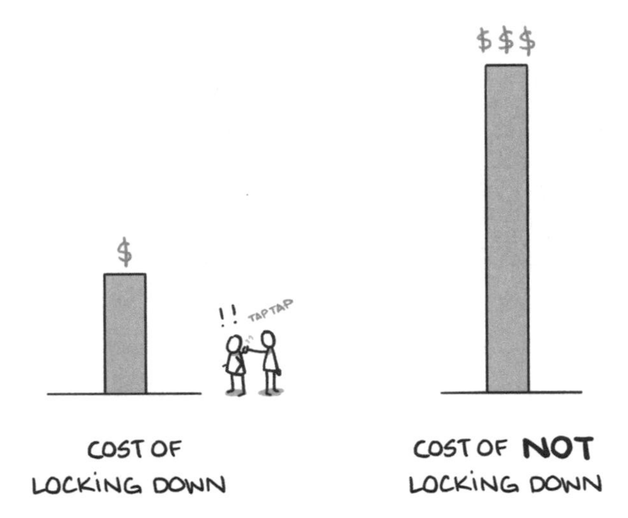

Long-term school closures might not be necessary in the future.

In the COVID era, if there's one issue that is debated almost as much as vaccines, it's whether to close schools.

Between March 2020 and June 2021, virtually every country in the world closed schools at some point because of COVID. The peak came in April 2020, when almost 95 percent of the world's schools had shut their doors. By June of the next year, all but 10 percent had reopened at least partially.

The arguments in favor of closing schools are compelling. Schools, with their constant interactions among kids, are already known as breeding grounds for the common cold and the flu—why would they be any different for some other pathogen? Teachers and staff aren't paid to risk their lives, which is exactly what the older ones are doing if they're required to teach in person without a vaccine during a pandemic like COVID. With this particular virus, the risk of severe illness or death goes up as you get older—an important factor to keep in mind when you think about how to allocate vaccines and other tools, a subject I'll return to later.

On the other hand, when schools closed, students fell behind in

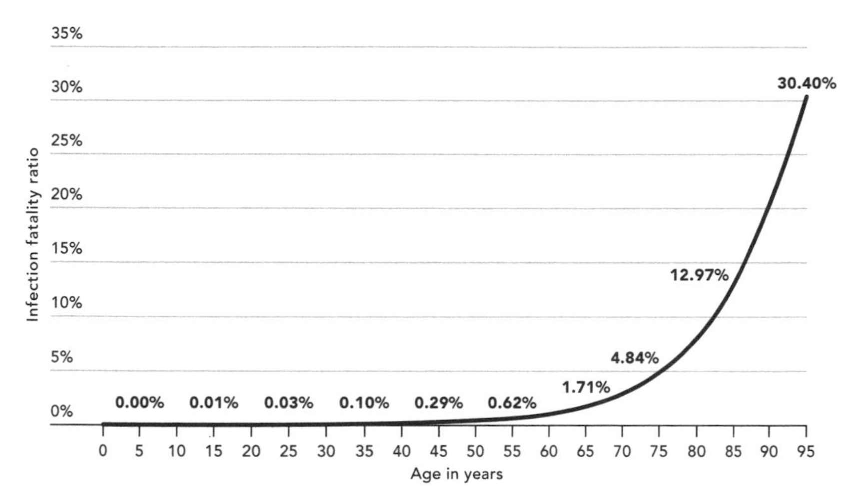

COVID is much worse for older people. This chart shows the estimated percentage of people infected with COVID who died in 2020. Notice how the curve shoots up for the elderly. (IHME)

their learning, and the achievement gap between wealthy and poor children grew even bigger than it already was. The United Nations estimates that COVID robbed students of so much time with their teachers that 100 million of them have fallen below the minimum threshold for basic skills, and it will take years of remedial work to help them catch up. In the United States, Black and Latino third graders have fallen twice as far behind in their classwork as white and Asian American students have. And the shift to remote school set white students back by one to three months in math, while students of color lost three to five months.

The pandemic also exposed one of the biggest myths about remote education—that it could ever replace classroom work for kids in the early grades. I'm a big fan of online learning, but I have always thought of it as a supplement to, not a substitute for, the work that young students and teachers do together in person. (In the United States, we mostly use the terms *remote learning* and *online learning* interchangeably, but many other countries provided lessons over the radio, television, and e-books as well as online.)

Few teachers have been trained to develop remote lessons, though this will change over time, as online tool kits and curricula improve. There are still many people who don't have internet access—in South Asia, more than a third of the students forced to stay home were unable to do remote learning—and many of those who do have access found the experience less than engaging. In short, online learning was put to a test that it was never designed to pass. I'm still optimistic about its future when it's used appropriately, though, and will have much more to say about it in the Afterword.

When schools close, the losses ripple out far beyond academic learning. Parents have to scramble for child care when their kids are suddenly at home during working hours. Millions of students in the United States and around the world rely on schools for free and reduced-price meals. In school, kids learn how to interact with their peers, get exercise, and have access to mental health support.

Unfortunately, the debate over closing schools has been muddied by some initial data that turned out to be misleading. Early on, there were fewer COVID cases among children, and a study in Norway found that there wasn't much transmission in schools, leading many people (including me) to conclude that kids weren't as susceptible as adults are. That was an argument, I thought, for keeping schools open.

But it wasn't true. In the United States through March 2021, the rates of infection and illness in children were comparable to the rates in adults ages eighteen to forty-nine, and even higher than the rates in adults ages fifty and older. The initial view was probably affected by the fact that many schools were closed; kids weren't less susceptible, they just had fewer opportunities to get infected. And when they did get infected, they were much less likely to show symptoms or get sick enough for their parents to get them tested—a problem that would have been rectified by large-scale testing.

Even with that in mind, what I think this all adds up to is that long-term school closures should not be necessary in future outbreaks, especially if the world meets the goal of producing enough vaccines for everyone within six months. Once vaccines are available, teachers should be near the front of the line for them (as many were when COVID vaccines first came out). If the disease is much worse for older people, as COVID is, you will probably want to distinguish between younger teachers and teachers who are older or who live with elderly people. (Remember that the age-related risks go down a lot for people under age fifty.) And in the meantime, many schools will be able to stay open while using layered prevention strategies, including masks, distancing, and better ventilation. One study found that reopening schools in Germany did not cause cases to go up, but reopening them in the U.S. did. The authors hypothesized that Germany's mitigation measures were more effective than America's.

I want to add a caveat to the idea that long-term school closures shouldn't be necessary. That will be true if the next outbreak is one with a profile like COVID's—in particular, one that rarely makes kids severely ill. But we have to be careful not to get caught fighting the last war. If a future pathogen is markedly different from COVID—if, for instance, its impact on children is a lot worse—then the risk/benefit calculus could change, and closing schools might be prudent. We'll need to stay flexible and, as always, follow the data.

On the other hand, it is clear to me that locking down assisted living homes for older people was the right thing to do. It saved a lot of lives because the virus is so much deadlier for the elderly—and I say this knowing how deeply painful and lonely these lockdowns were for all the residents who were confined to their rooms, and for their loved ones. It was heartbreaking to hear stories about families who had to say goodbye to a dying parent or grandparent through a closed window or over the phone. My dad died of Alzheimer's disease in September 2020, and I feel very fortunate that he was able to be at home, surrounded by family, in his last days.

The human suffering caused by these separations is incalculable, literally—no one can put a number on the pain of not being able to say goodbye in person. But the policy saved so many lives that it will be worth adopting again if the circumstances call for it.

What works in one place might not work in another.

No matter where you are in the world, a mask is going to give you the same protection. Unfortunately, many other NPIs aren't so universal. Their effectiveness depends a lot not only on when they're used, but on where they're used too.

Lockdowns are a great example. The evidence is clear that they reduce transmission, and that stricter lockdowns reduce transmission more than looser ones do. But they're not equally effective everywhere, because not everyone is able to comply by staying in one place.

The difference is actually quantifiable. One ingenious study used anonymous cell phone data from throughout the United States to gauge the extent to which people living in different neighborhoods stayed at home. (Your cell phone checks in periodically with a service that establishes its location.)

Between January and March 2020, people who live in America's wealthiest neighborhoods were the most mobile—that is, they spent the most time out of their homes—and people in the lowest-income neighborhoods were the least mobile.

But in March, as lockdowns kicked in across the country, the situation swung upside down. People in wealthy neighborhoods became the least mobile, and people in the poorest ones were the most mobile. The reason: They were much less likely to have jobs they could do from home, and much less able to use delivery services for groceries.

A similar shift was driven by population density. Before lockdowns, the densest communities had the highest transmission rates. After lockdowns, they had the lowest transmission rates, and the less-crowded areas had not dropped nearly as much. That makes sense, of course, because when people aren't living and working in close quarters to begin with, telling them to stay home will naturally have less of an impact on transmission.

Researchers have drawn other conclusions about the differences among and within countries. Contact tracing is more effective in places with a good system for reporting and processing the data about each person's contacts, though once the case numbers get high, it becomes much harder. Social distancing and lockdowns work better in richer countries than in poorer ones, for many of the same reasons that explain why they work better in wealthier parts of the U.S. than in poorer parts. In some countries, lockdowns can backfire, as the disease is spread by people who migrate (returning to their home community from a job in the city, for example). Lockdowns may not be necessary in places where the disease burden is modest. They're also more effective in countries where residents have less of a voice in the country's affairs, and the government is in a position to strictly enforce lockdowns and other mandates.

What this all means is that there is no single ideal mix of NPIs that works equally well everywhere. Context matters, and protective measures need to be tailored for the places where they'll be used.

The flu almost went away, at least for a while.

In the fall of 2020, as we got closer to flu season, I started to worry. Every year, influenza kills tens of thousands of Americans and hundreds of thousands of people around the world, nearly all of them elderly.* Even more are hospitalized. At a time when COVID was overwhelming or at least sorely testing virtually every health system on the planet, a bad flu season could have been disastrous.

But there was not a bad flu season that year. In fact, there was hardly any flu season at all. Between the flu seasons of 2019–20 and 2020–21, cases dropped 99 percent. As of late 2021, one particular type of flu known as B/Yamagata had not been detected anywhere in the world since April 2020. Other respiratory viruses also dropped dramatically.

By the time you read this book, of course, things may have changed. Flu strains have a way of disappearing for long periods and then suddenly recurring without explanation. But the huge decline in cases across the board is unmistakable, however long it lasts, and we know why: Nonpharmaceutical interventions made a dramatic difference in reducing flu transmission when combined with the prior immunity and vaccinations that people had.

This is great news, and not only because it means we did not have a disastrous twindemic of flu and COVID in 2020–21. It's also reason to be hopeful that if there's an outbreak of a bad flu in the future, NPIs can help keep it from turning into a pandemic. While it's possible that we'll see an influenza so transmissible that it overwhelms our best efforts to contain it without an updated vaccine, it is reassuring to have more evidence that NPIs are effective against the common strains we know about today. And we now have solid evidence that, when paired with vaccines, NPIs could help us eventually eradicate every strain of flu.

* The estimates of how many people get sick and how many die from the flu every year vary widely. Deaths in particular are probably undercounted, because not all flu deaths get reported to infectious-disease centers like the CDC, and because flulike symptoms might not be reported on a death certificate.

We should use contact tracing to find superspreaders.

Depending on which country you live in, if you tested positive for COVID, you may have been called by someone asking about all the people you might have come into contact with. They probably focused specifically on the forty-eight hours before you first started feeling sick (if you did feel sick). This is the process known as contact tracing.

Although it felt new to many people around the world during COVID, contact tracing is actually an old strategy. It was essential to eradicating smallpox in the twentieth century and is also at the center of strategies to fight Ebola, tuberculosis, and HIV in the twenty-first.

Contact tracing works best in countries that excel at testing and processing data—among them, South Korea and Vietnam. But both of those countries did things that wouldn't fly in the United States. Under a law changed after the 2014 outbreak of Middle East Respiratory Syndrome, or MERS, the South Korean government used data from credit cards, mobile phones, and surveillance cameras to trace the movements of infected people and identify other people they had come in contact with. It published this information online, though it had to restrict some of the data after regional governments gave out too many details about people's movements. According to the journal *Nature,* one man "was wrongly accused of having an affair with his sister-in-law because their overlapping maps revealed they dined together at a restaurant."

Vietnam also used posts on Facebook and Instagram, along with mobile phone location data, to supplement extensive face-to-face interviews. In March 2020, before the country was testing every passenger arriving from the U.K., a flight from London arrived in Hanoi with 217 passengers and crew. Four days later, a patient went to the hospital with symptoms and tested positive for COVID. Vietnamese authorities tracked down all 217 people from the flight and identified 16 more cases among them. Everyone on the plane and more than 1,300 of their contacts were quarantined. In all, there were 32 cases related to the flight, a small fraction of the cases that would've happened if all those passengers and crew members had simply gone on their way.

If you read the previous two paragraphs and thought, *If anyone calls me about contact tracing, I'm not picking up the phone,* you're not alone. In two counties in North Carolina, many of the contacts who were named never returned the tracer's phone call. And between a third and half of COVID-infected people who were reached claimed not to have had contact with a single person in the days before they tested positive. But contact tracing will often be an important part of stopping the spread of a disease, which is why we have to figure out how to build trust between public health agencies and the public so that more people will share their contacts.

One of the reasons people hesitate to respond is the fear that their contacts will have to go into quarantine, but, fortunately, broad quarantines of every single contact might not always be necessary. In England, some schools had their students stay home for ten days if they had come into contact with someone who had COVID. Other schools allowed kids to keep coming to class as long as they tested negative every day. This daily testing, it turns out, was just as good at preventing outbreaks, but without having to keep students home from school.

And contact tracing can still be effective even if it's not done as intensively as Vietnam and South Korea did it. In general, if you start the program when only a small fraction of the population is infected, and if you identify a large share of cases in your country, contact tracing can cut transmission by more than half.

Some U.S. states and other governments rolled out smartphone apps that helped identify possible contacts, but I'm skeptical that these apps will ever be effective enough to be worth a big investment of money or time. For one thing, their usefulness is capped by the number of people who install them, because the apps only record an exposure if both parties who come into contact with each other are using them. I suspect that most of the people who use these apps are also the ones who would follow lockdown guidelines—and if you're doing that, you should have so few contacts that you can probably remember them all. For the people who really are sheltering in place, getting a message that says "Hey, you saw your brother" won't be very helpful.

During COVID, one challenge with conventional contact tracing is that it's not an especially efficient use of resources, because the virus is not transmitted at the same rate by everyone who's infected. If you get the original COVID strain, the chances are not especially high that you'll pass it along to someone else. (About 70 percent of those cases may not transmit to anyone else at all.) But if you *do* pass it along to someone else, you probably pass it along to many people. For reasons we don't entirely understand, 80 percent of COVID infections with early variants came from just 10 percent of the cases. (These numbers could be different for the Omicron variant—as I write this, we don't have enough data to know.)

So with a virus like COVID, using the conventional approach means you'll spend a lot of time finding people who wouldn't have infected anyone else—epidemiologically speaking, you'll find yourself in a lot of cul-de-sacs. What you really want to do is find the main thoroughfares, the relatively small number of people who are causing the most infections.

Understanding this limitation, a few countries tried a newer approach to contact tracing. Instead of working forward in order to find out whom they might have infected, these countries worked backward—identifying contacts up to fourteen days before the person started feeling sick. The goal was to find out who may have infected the patient, and then see who else that person may have passed the virus to.

Backward contact tracing is hard to pull off unless you have widespread testing, fast results, and a system for contacting people quickly, and it's especially hard when you're dealing with a pathogen that spreads fast, because there's not much time between getting infected and becoming contagious. But where it was practical, the approach worked really well. It was used at locations in Japan, Australia, and other countries, and it proved quite effective at finding individuals who were superspreading the early COVID variants. One study found that it could prevent two to three times more cases than the traditional approach.

It's stunning how little we know about superspreaders. What role does biology play? Are some people more prone to being a superspreader than others? There's also certainly a behavioral component. Superspreaders seem not to pose more of a risk to small groups than other infected people do, but in crowded indoor public spaces, such as bars and restaurants, there's a better chance that you'll encounter one or more superspreaders, and they'll have the opportunity to infect a lot of people. Superspreaders are one of the mysteries of disease transmission that needs a lot more study.

Good ventilation matters more than you might think.

Remember the advice early on to wash your hands a lot and avoid touching your face? Or the way cashiers would clean a pen each time someone used it to sign a credit card receipt? Or how you'd feel safe if you stood far enough away from someone you were talking to?

It is a good idea to wash your hands, clean pens, and keep your distance: They're generally good health practices, and they help keep other pathogens like the flu and the common cold at bay. And it's clear that soap and disinfectants do break down the COVID virus, rendering it harmless.

But after two years of COVID, scientists know a lot more than they did in early 2020 about how this particular virus spreads. One finding stands out: The virus can linger in the air longer and travel farther than most people thought in early 2020.

You may have heard some of the anecdotal evidence. In Sydney, Australia, an eighteen-year-old man singing in a church loft passed the virus to twelve others who were sitting fifty feet away. At a restaurant in Guangzhou, China, a single person infected nine others, including some who were sitting at the same table but also some who were seated at tables several feet away. In Christchurch, New Zealand, someone staying in a quarantine hotel picked up the virus through an open door nearly one minute after an infected person had walked past the room.

None of this is speculation. Researchers looking at these cases have rigorously ruled out all the other ways the infection could have spread. One group of scientists who studied the Guangzhou incident used video footage to count thousands of times that waiters and customers in the restaurant touched the same surfaces; the number wasn't nearly high enough to explain all the cases. The case in New Zealand is supported by genetic analysis: By studying the genomes of the virus in both infected people, scientists determined that the second patient almost certainly got it from the person who walked past.

The good news is that COVID's airborne proclivities could be a lot worse. It appears that the virus is able to stay in the air for several seconds, and perhaps several minutes. The virus that causes measles, on the other hand, can stay in the air for hours.

To understand why viruses get transmitted through the air, we need to talk about your breath.

Whenever you talk, laugh, cough, sing, or simply breathe out, you exhale. We tend to think of the stuff we're exhaling as air, but it contains a lot more than that. Your breath is filled with tiny globs of liquid, a mix of mucus, saliva, and other secretions from your respiratory tract.

These globs are grouped into two categories by size: The larger ones are known as droplets, and the smaller ones are called aerosols (not to be confused with the air fresheners and hairsprays that come in a can). The dividing line between them is typically 5 micrometers, which is roughly the size of the average bacterium. Anything bigger than that is a droplet, and anything smaller is an aerosol.

Droplets, being on the larger side, typically contain more virus than an aerosol, which makes them a better mechanism for transmission. On the other hand, because they're relatively heavy, they don't make it more than a few feet from your mouth or nose before falling to the ground.

The surface that a droplet lands on becomes what's called a fomite, and how long the fomite is able to transmit the virus depends on several factors, including the type of pathogen and whether you sneezed or coughed it out (in which case it's more protected because it's covered in your mucus). Studies show that even though the COVID virus may be able to survive for a few hours, or even days, it's quite rare for people to get sick from touching a contaminated surface. In fact, even if someone does happen to touch a fomite, the chances that the person will get infected are less than 1 in 10,000.

Once it was known that COVID spread mainly through the air, most experts thought it did so via droplets. This would have meant that anyone more than a few feet away, or sharing the same air space just a few seconds later, would've been safe. But further research showed that aerosols also contribute significantly to transmission. They're capable of containing quite a bit of virus, and because they weigh much less than droplets, they can travel farther and linger in the air longer. And for a time, at least, the virus was evolving to spread even more via aerosols—people with the Alpha variant exhale about eighteen times more virus in aerosols than people with the original virus do.

Part of the reason aerosols were underestimated is that, because they're so small, they usually dry out quickly, which makes the virus particle inactive. One study used a computer simulation to show that COVID viruses—particularly the Delta and Omicron variants—have an electrical charge that attracts substances from the lungs which slow down the process of drying out aerosols. We need to study transmission dynamics a lot more so that next time we understand quickly how transmission is taking place.

Depending on the conditions in the room—temperature, airflow, humidity—aerosols that contain COVID virus may be able to travel several feet. It's not clear yet what share of cases is caused by aerosol transmission, but it could be more than half.

What does this all add up to? Airflow and ventilation matter, probably quite a bit. You should install high-quality air filters to remove aerosols if you can, and if that's not possible, there's also a simpler and cheaper option: opening the windows. In a study done in Georgia, schools that opened doors or windows and used fans to dilute airborne particles had around 30 percent fewer COVID cases than those that didn't. Schools that also installed air filters had 50 percent fewer cases.

It's good to wash your hands and wipe down surfaces, and in a future outbreak, that might be your first choice for staying safe. But when it comes to preventing COVID, if you have to choose between spending time and money on cleaning things or improving airflow, improve the airflow.

Social distancing works, but there's nothing magical about six feet.

I've lost count of how many signs I've seen reminding me to stay six feet away from others. My favorite is the one at the club where I play tennis, which hilariously explains that six feet is the width of twenty-eight tennis balls. How many people in the world are so tennis-oriented that they understand twenty-eight tennis ball widths better than they understand six feet? If you get too close, do they say, "Hey, you're only nineteen tennis balls away, please step back nine more?" I suppose that if such people exist, you'll find them at a tennis court. But I play a lot of tennis and I have no clue what a distance of twenty-eight tennis balls looks like.

In any event, there's nothing magical about the six-foot rule (or the twenty-eight-tennis-balls rule). The WHO and many other countries recommend distancing of 1 meter, or about 3 feet. Others recommend 1.5 to 2 meters, which is around 6½ feet.

In truth, there is not some stark cutoff where you're at high risk of getting COVID within a certain distance and zero risk beyond that distance. There's a continuum of risk, and it depends on the specific situation you're in: how big the droplets you're exposed to are, whether you're inside or outside, and so on. Six feet is better than shorter distances, but we don't know just how much better. Before the next pandemic, scientists need to dig into this question and help us understand the role of ventilation and air movement so we can get a clearer answer.

In the meantime, the six-foot rule is a good one to follow, unless it's very difficult to maintain, such as in a classroom. People need clear, easily remembered guidelines. It is not a helpful public health message to say, "Keep your distance, but the exact distance depends on the situation, so it might be three feet, or six feet, or maybe more."

It's mind-blowing how cheap and effective masks are.

This is a little hard to admit, because the power of inventing things is so central to my worldview, but it's true: We may never devise a cheaper, more effective way to block the transmission of certain respiratory viruses than a piece of inexpensive material with a couple of elastic straps sewn onto it.

The idea of controlling a disease by promoting the widespread use of masks is both simple and old. It dates from 1910, when Chinese authorities called on a pioneering physician named Wu Lien-teh to lead the response to an outbreak of pneumonic plague in the region then known as Manchuria, in the northeast part of the country. The disease had a fatality rate of 100 percent—every single infected person died, sometimes within twenty-four hours—and was thought to be spread by infected fleas that lived on rats.

Wu believed that it was airborne, not rodent-borne, and he insisted that medical staff, patients, and even members of the general public cover their faces with masks. He was partly right—you can get the infection from a flea carried by rats, but the more dangerous situation is when people have plague in their lungs and then transmit it through the air to other humans. Although 60,000 people died before the outbreak had run its course, the consensus was that Wu's strategy had prevented it from being much worse. He was celebrated as a national hero, and thanks in large part to his leadership, masks became a common sight—as protection against disease, air pollution, or both—throughout China. Even if COVID had not happened, they would still be part of the country's social fabric today.

Just as Chinese experts were initially mistaken about how the 1910 plague was transmitted, much of the scientific community in the West was initially wrong about how COVID spread. ("The big mistake in the U.S. and Europe," the head of the Chinese CDC said in March 2020, "is that people aren't wearing masks.")

For many people who were following the research—people in the United States, at least—the argument in favor of masks was settled by an incident involving two hairstylists at a salon in Springfield, Missouri.

Both stylists developed symptoms and tested positive for COVID in May 2020. Their records indicated that they had exposed 139 clients. But everyone had masked up during the haircuts, and not a single client developed symptoms.

Was that because the stylists weren't transmitting the virus? No. One of them had four close contacts outside the salon—when the stylist wasn't wearing a mask—who developed symptoms and tested positive. That settled the question. Like a good set of barber shears, masks were cutting transmission.

The Springfield incident shows how masks can actually serve two purposes: to prevent an infected person from spreading the disease, and to protect an uninfected person from getting it. The first is called source control, and the beauty of it is that almost any mask of any type helps with source control, at least for many viruses. Both cloth and surgical masks prevent about 50 percent of particles from escaping when you cough, and when they're worn together, they can block more than 85 percent.

The second purpose of masks—protecting someone from getting infected—is a little more challenging if they don't fit tightly. According to one study, if you're wearing a loose surgical mask and sitting six feet away from an unmasked person with COVID, your mask will reduce your exposure by only 8 percent. Double masking helps a lot, reducing your exposure by 83 percent.

The real benefit comes with universal masking, where both people are double masking or improving the fit of their surgical masks: It reduces the risk of exposure by 96 percent. That's an incredibly effective intervention that can be manufactured for just a few cents.

(By the way, some of the experiments used to test this kind of thing are wonderfully creative. One research team padded the inside of a mannequin's head to simulate the nasal cavities of a human skull, set it at a height of 5 feet 8 inches—near the global median height for men—and hooked it up to a smoke machine and a pump. Then they measured how far the droplets traveled when the mannequin coughed in various scenarios: with its mouth uncovered, with its mouth covered by a bandanna made from a T-shirt, with a folded handkerchief, and finally with a stitched mask. Another group of researchers set two mannequins next to each other, simulated a cough from one of them, and then measured how many particles made it from the cougher to the coughee.)

The reason double masking works so well is that it forces the masks to fit more snugly on your face. Some higher-quality N95 or KN95 masks, called respirators, are designed to do that by themselves.* One study found that properly fitted respirators were 75 times more effective than well-fitted surgical masks, and even loose respirators were 2.5 times better than well-fitted surgical

* In other parts of the world, an equivalent respirator might be called an FFP2, a KF94, or a P2.

masks. (In case you're wondering, the 95 indicates that, in testing, the mask's material blocked 95 percent of very small particles blown with enough force to simulate a human doing hard work. On an N mask, the elastic straps go around the back of your head, while the straps on a KN mask go around your ears.)

Respirators like the KN95 (left) do the best job of protecting you and people around you, especially from highly transmissible viruses. Surgical masks (middle) and cloth masks (right) are also highly effective, especially when everyone wears them.

Early on in the pandemic, when hospitals and clinics were running out of respirators, it was important to reserve the limited supply for health care workers who were putting themselves at risk to treat patients. But as I write this, two years since the first cases were identified, the supply isn't constrained anymore—so there's no good reason that respirators still aren't readily available to everyone in America. (Some countries, like Germany, require you to wear them in public spaces.) This became a bigger problem as COVID evolved to become more transmissible—every chain is only as strong as its weakest link, and masks can only stem an outbreak if enough people wear them.

In America, unfortunately, resistance to wearing masks is almost as old as masks themselves. During the 1918 flu pandemic, just a few years after Wu's breakthrough finding, several U.S. cities adopted mask mandates. In San Francisco, anyone who failed to wear a mask in public could be fined or put in jail. Protests broke out across the city. In October 1918, a "mask slacker" beat a health inspector with a bag of silver dollars for insisting that he cover up. The health inspector pulled out a pistol and shot him.*

It is unfortunate that Americans have not grown more accepting of masks in the intervening century. The protests were at least as vocal and occasionally violent in 2020 as they were in 1918.

Missing the value of masks is indeed, as the head of China's CDC said, one of the biggest mistakes made during the pandemic. If everyone had masked up early on—and if the world had had enough supplies to meet the demand—it would have blunted the spread of COVID dramatically. As a public health expert told me over dinner one night, "If everyone would just wear masks, *How to Prevent the Next Pandemic* would be a very short book."

The benefit of masks has now been proven around the world. Early on in the pandemic, Japan took masking seriously, which in combination with the country's backward contact tracing kept its excess deaths at an extraordinarily low 70 per million people at the end of 2021. (Remember that the U.S. was around 3,200 per million at that point.) And in Bangladesh, researchers ran a study involving nearly 350,000 adults in 600 villages to look at the impact of public messaging about masks. One group, about half of the people, were given free masks (some cloth, some surgical), information on the importance of using them, in-person reminders, and encouragement from religious and political leaders. The second group didn't get any of those things. After two months, proper mask use in the first group

* Both men survived. According to *The New York Times,* the "slacker" "was charged with disturbing the peace, resisting an officer and assault. The inspector was charged with assault with a deadly weapon."

was up to 42 percent, versus just 13 percent in the second group. People in the first group also had a lower COVID infection rate, and even five months later, they were still more likely to be using masks.

All of this can get a bit complicated, but the key thing to remember is that masks work. Cloth and surgical masks are highly effective, especially when everyone wears them. In high-risk settings and with highly transmissible viruses, respirators are even better. In any case, masks and respirators are dirt cheap and more effective than any vaccine or drug that we have so far.

It'll be interesting to see whether the social standards on wearing masks change much as a result of COVID. In March 2020, I went to an in-person meeting while I was feeling under the weather. Because the CDC hadn't recommended masks yet, I didn't wear one. Luckily, I later found out that I'd had the flu and not COVID, but I feel bad that I was there with respiratory symptoms without taking a measure that might have reduced the chance of spread. Knowing what I know now, I would either join that meeting virtually or wear a mask.

But will that practice catch on more widely? It's hard to say. My guess is that most Americans will eventually go back to attending meetings and big sporting events without masks. So we'll need to get the word out about masking up if you're experiencing respiratory symptoms, and we'll need the public warning systems to kick into high gear as soon as there's a sign of a problem. It could make the difference between an outbreak and a pandemic.

CHAPTER 5

FIND NEW TREATMENTS FAST

Early on, rumors and misinformation about COVID seemed to be spreading faster than the disease itself. In February 2020, a month before it had declared COVID a pandemic, the WHO was already contending with false claims about various substances that supposedly cured or prevented the disease. Its director-general said, "We're not just fighting an epidemic; we're fighting an info-demic," and its website began featuring a myth-busting section that had to be constantly updated in order to debunk false claims.

In just the first half of 2020 alone, doctors had to shoot down false rumors that COVID could be cured by

- Black pepper
- Antibiotics (COVID is caused by a virus, and antibiotics don't affect viruses)
- Vitamin and mineral supplements
- Hydroxychloroquine
- Vodka
- Sweet wormwood

Although none of these substances has any effect on COVID, I can see why people would want to believe otherwise. Some of them are legitimate medical interventions: Hydroxychloroquine is used to treat malaria, lupus, and other diseases, and ivermectin is a standard treatment for various parasitic diseases in people and other animals. Obviously, just because a drug treats one condition doesn't mean it will work on COVID, but it's not irrational to hope that it might.

I can even see why people might be drawn to purported cures that are closer to a folk remedy than to modern medicine. At a time when a frightening new disease is making its way around the world and our phones are sending us the latest scary stories about it by the day or even by the hour, it's natural to look for immediate help anywhere you can find it. Especially when there's no scientifically proven cure to fill the need for a treatment, and when the alternative being proposed is already in your bathroom cabinet or under your kitchen sink.

There is nothing new, of course, about people clinging to the false hope of an easy cure. Humans probably started doing it as soon as they were aware enough of their own mortality to look for ways to fend it off. But medical misinformation is more dangerous now than ever, because it can travel faster and farther than ever, with tragic consequences for many of the people who believe it.

I don't know of a complete solution to this problem. But I do think there would have been fewer wrong ideas going around about COVID if science had found an actual treatment sooner—something that everyone could point to as a legitimate therapeutic—and if it had become widely available around the world.

Early on in COVID, that's what I thought would happen. I was confident that a vaccine would be developed eventually, but I expected that a treatment would come along well before that point. I wasn't alone: Most people I know in the public health community felt the same way.

Unfortunately, that's not what happened. Safe, effective COVID vaccines were available within a year—a historic feat that will get its due in the next chapter—but treatments that could keep large numbers of people out of the hospital were surprisingly slow out of the gate.

It wasn't for lack of trying. Doctors began prescribing hydroxychloroquine off-label—that is, for something other than its approved purpose—almost from Day 1. Early reports suggested that it could be effective against COVID, and the FDA gave the provisional green light known as emergency-use authorization.

The early evidence for hydroxychloroquine came from lab studies of its effect on cells taken from the kidneys of the African green monkey. These cells are often used to screen potential antiviral drugs because viruses replicate very quickly in them, and in fact the method did surface some promising treatments, including the antiviral drug remdesivir.

In early studies, hydroxychloroquine was able to block one pathway by which the COVID virus entered the monkey cells, suggesting that it might do the same in humans. Hundreds of clinical studies tried to replicate these promising results, but in early June an authoritative randomized study in the U.K. found that the drug didn't provide any benefit for patients who'd been hospitalized with COVID. Ten days later, the FDA revoked the emergency-use authorization, and two days after that the WHO dropped hydroxychloroquine from a trial it was running.

The problem, it turns out, is that human cells have a different pathway from the one that the drug blocked in monkey cells—so the promising results didn't translate from animals to people. As far as treating COVID goes, the drug was a dead end. Meanwhile, the hydroxychloroquine craze caused a run on the drug, and many patients who needed it to treat lupus and other chronic conditions couldn't get it.

By that summer, dexamethasone had become the primary treatment for severe COVID, having been found to reduce mortality among hospitalized patients by nearly a third. Dexamethasone, a steroid that has been in use since the 1950s, works on COVID somewhat counterintuitively: by suppressing some of the immune system's defenses.

Why would you want to suppress your immune system? Because once you're past the early stages of infection, the biggest danger of COVID doesn't actually come from the virus, it comes from the immune system's reaction to the virus.

In most people, the immune system is able to reduce the amount of virus in the body within five or six days of getting infected. But then it becomes so activated that it can cause an intense inflammatory phenomenon known as a cytokine storm—a flood of signals that causes blood vessels to leak massive amounts of fluid into various vital organs. (With COVID, this leaking is a particular problem in the lungs.) This intravascular loss of fluid can also lead to dangerously low blood pressure, which in turn can cause further organ failure and death. It's your body's overreaction to the invasion that makes you sick.

Dexamethasone was a significant success: It was effective, easy to deliver, cheaper than any of the alternatives, and widely available even in many developing countries. (In fact, even before COVID, the WHO deemed it an essential medicine for use in pregnant women.) Less than a month after it was shown to work well, the African Medical Supplies Platform—the group that distributed LumiraDx testing machines to African countries—had acquired enough tablets to treat nearly one million people throughout the African Union, while UNICEF made an advance purchase to treat 4.5 million patients. British researchers estimated that by March 2021, dexamethasone had saved as many as one million lives around the world.

Even so, the drug does have its downsides—chiefly that, if used too early, it will mute your immune response at the very moment when it needs to be at full strength so it can stop the virus from replicating. When that happens, you become more susceptible to complications and opportunistic infections. The second wave of COVID in India was accompanied by a spike in cases of a gruesome and deadly disease called mucormycosis, also known as "black fungus"—some people had this fungus in their lungs, but it was held in check until their immune system was suppressed, unleashing it and causing the disease. In most countries, almost no one had this fungus, so the problem was mostly limited to India.

Hoping to find another existing drug that might help, researchers tried dozens of other potential treatments that were already at hand. For example, there are various ways to take antibodies from the blood of people who have recovered from a disease and give them directly to someone who's still sick, an approach known as giving convalescent plasma. Unfortunately, this approach wasn't effective or practical enough to warrant its broad use for COVID. Remdesivir, the antiviral that showed promise in monkey cells, was originally developed to fight off hepatitis C and RSV, and early studies showed that it didn't help hospitalized patients enough to be worth ramping up for more people. (It was also hard to administer: It required five daily infusions!) However, a subsequent study showed that it may have a major impact in patients who aren't yet sick enough to be in the hospital, demonstrating that sometimes a product can find its niche if it reaches the right people at the right time. Even so, remdesivir needs to be given intravenously for three days early on in the course of the disease, so it will be important to find a modified form that can be delivered by inhaling it or taking it as a pill.

Although convalescent plasma didn't pan out for COVID, I hoped we would have more luck with a different approach to giving people antibodies. It's called monoclonal antibodies, or mAbs for short, and it worked well enough to receive an emergency use authorization for COVID cases in November 2020—only a month before the first vaccines became available.

Instead of preventing a virus from taking over healthy cells, or from replicating once it does take over a cell—which is how most antiviral drugs work—mAbs are identical to some of the antibodies that your immune system generates to mop up the virus.

(Antibodies are proteins with variable regions that allow them to grab onto unique shapes on the surface of the virus.) To make mAbs, scientists either isolate a powerful antibody from a person's blood or use software modeling to come up with an antibody that grabs the virus. Then they clone it billions of times. This cloning from a single antibody is why they're called monoclonal.

If you're infected with COVID and get mAbs at the right time (and if they're adapted for the variant you have), they reduce the risk that you'll end up in the hospital by at least 70 percent. I had high hopes for mAbs in the early days of COVID—so much so that the Gates Foundation paid to have up to 3 million doses set aside for high-risk patients in poor countries. But we soon learned that mAbs weren't going to be a game changer for COVID: The Beta variant of the virus, which was especially prevalent in Africa, had changed its shape enough that the antibodies we had supported no longer grabbed onto it enough to help. We could have started over to develop another mAb that would have been effective for the new variant, but manufacturing it would have taken three to four months, which would have made it hard to keep up with a virus that evolves as quickly as COVID does.

In the future, there may be better ways of manufacturing mAbs that reduce this lead time so we can get them out quickly and cheaply. And we should be looking for mAbs that grab onto a piece of the virus that is unlikely to change. As I write this, a mAb called Sotrovimab, which was isolated from a SARS patient and then modified, has shown to be broadly effective against all the known COVID variants—giving us reason to hope that scientists will be able to create antibodies that work on broad families of viruses.

Other downsides became clear as wealthier countries tried to roll out mAbs treatments. COVID antibodies were expensive to make, they had to be administered at facilities capable of infusing them into your blood, and they helped only those patients who could be identified early in the course of the disease. The lack of facilities was a particularly big problem in developing countries. Because of these problems, we wrote off our investment in mAbs for COVID—though we're still supporting a lot of work on mAbs for other diseases—and increased our focus on antiviral drugs, particularly ones that patients could take orally rather than intravenously.

As soon as COVID was identified, many researchers started looking for the holy grail of treatments: an antiviral drug that's cheap, easy to administer, effective for different variants, and capable of helping people before they get terribly sick. In late 2021, a few of these efforts paid off—not as soon as would have been ideal, but still in time to have a big impact.

Merck and its partners developed a new antiviral called molnupiravir, which could be taken orally and was shown to significantly reduce the risk of hospitalization or death for people at high risk. In fact, the drug worked well enough that the clinical trial was stopped early. (This is a common practice in trials—they'll wrap up early if it would be unethical to continue because there's definitive evidence either that the drug is a success, in which case the participants who aren't getting it are receiving clearly inferior treatment, or that the drug is a failure, in which case the participants who are getting it are the ones receiving inferior treatment.)

Soon the study of a second oral antiviral, Paxlovid (made by Pfizer), was also stopped because the drug worked so well. When Paxlovid was administered to high-risk patients soon after symptoms kicked in, and in combination with a drug that prolonged its effects, it reduced the risk of severe illness or death by nearly 90 percent.

By the time these announcements came out in late 2021, a large share of the world's population had received at least one dose of a vaccine. But that's no reason to think that therapeutics aren't important, in COVID or any other outbreak. It's a mistake to think of vaccines as the star of the show and therapeutics as the opening act you would just as soon skip.

Consider the time line. In the next epidemic, even if the world is able to develop a vaccine for a new pathogen in 100 days, it will still take a lot of time to get the vaccine to most of the population. This is especially true if you need two or more doses for full and continued protection. If the pathogen is especially transmissible and deadly, tens of thousands or more could die without a therapeutic drug.

Depending on the pathogen, we may also need ways to treat its long-term effects. For example, months after being infected with COVID, some people continue to experience terrible symptoms: difficulty breathing, fatigue, headaches, anxiety, depression, and the cognitive problems that became known as "brain fog." COVID isn't the first condition with long-lasting effects like these; some scientists have argued that similar symptoms can also be associated with other viral infections, trauma, or a stay in an intensive care unit. Still, researchers have documented that even a mild case of COVID can cause inflammation for weeks afterward, and that its impact isn't limited to your lungs—it can affect your nervous system and blood vessels too. We need to know a lot more about long COVID, as this condition is known, so we can help the people who are experiencing it now, and if the next major outbreak has a similar long tail, we'll need ways to treat those symptoms too.

Even once there is a vaccine, we'll still need good therapeutics. As we've seen with COVID, not everyone who can take a vaccine will choose to do so. Unless the vaccine completely prevents breakthrough cases, some vaccinated people will still get sick. If a variant comes along that the vaccine doesn't protect you from, we'll want to have treatments on hand until the vaccine can be tweaked. And, along with nonpharmaceutical interventions, therapeutics can reduce the strain on hospitals, which would prevent the overcrowding that ultimately means that some patients die who otherwise wouldn't.

With good enough therapeutics, the risk of severe illness and death will drop (in some cases dramatically), and countries can decide to loosen restrictions on schools and businesses, reducing the disruption to education and the economy.

What's more, imagine how people's lives will change if we're able to take the next step by linking testing and treatment. Anyone with early symptoms that might indicate COVID (or any other pandemic virus) could walk into a pharmacy or clinic anywhere in the world, get tested, and, if positive, walk out with a pack of antivirals to take at home. If supplies were short, the people with serious risk factors would get priority.

All of which is to say: Therapeutics are fundamentally important in an outbreak. We're lucky that scientists made COVID vaccines as quickly as they did—if they hadn't, and considering the slow progress toward effective treatments in the first two years of the pandemic, the death toll from COVID would have been far worse.

To understand how we can avoid what happened with COVID, we need to take a tour through the world of therapeutics: what they are, how they get from the lab to the market, why they didn't fare better early in this pandemic, and how innovation can set the stage for a better response in the future.

We tend to think of medicines as mysterious and complex substances, but the most basic ones are remarkably simple—clusters of carbon, hydrogen, oxygen, and other elements that can be described using nothing more than high school chemistry. Just as water is H_2O and salt is NaCl, the formula for aspirin is $C_9H_8O_4$. Tylenol is $C_8H_9NO_2$. Because these molecules have a very small mass, they belong to a class of drugs known as small molecules.

Small-molecule drugs have several advantages that make them especially appealing in an outbreak. Because their chemical structure is fairly straightforward, they're easy to manufacture, and thanks to their size and chemistry, they don't get broken down by your digestive system, so you can take them as a pill. (This is why you've never had to get an injection of aspirin.) And most of them can be kept at room temperature and have a long shelf life.

Larger molecules are more complicated in just about every respect. A monoclonal antibody, for example, is 100,000 times larger than a molecule of aspirin. Because large molecules are broken down by your digestive system if you swallow them, they need to be injected or given by an IV drip. This means you'll need medical personnel and equipment to make sure they are administered properly, and you'll need to isolate infected patients when they come in for treatment so they don't pass the virus to other people at the facility. Large molecules also require far more complex manufacturing—they're made using live cells—which means they're more expensive, and it takes more time to produce them at high volume.

In short, during an outbreak, you'd rather have small-molecule treatments than large ones, all other things being equal. But we may not be able to find a small-molecule drug that works well against a particular pathogen (or that works without causing bad side effects), so our pandemic plan should have us ready to pursue small- and large-molecule treatments in parallel. Over the next decade, we can do research and development to shorten the steps required and reduce the cost of manufacturing when a potential pandemic is detected.

We'll also need to deliver other lifesaving tools that aren't drugs but do help patients stay alive long enough for their bodies to recover. Oxygen is a prime example: According to the WHO, in early 2021 about 15 percent of COVID patients got so sick that they needed supplemental oxygen.

Oxygen is an important component in any health system—it's used in cases of pneumonia and premature birth, among others—and although rich countries have run short of it during COVID, low- and middle-income countries have struggled even more. One survey found that only 15 percent of health facilities in developing countries had any oxygen equipment, and only half of that equipment was functional. Hundreds of thousands of people die every year because they can't get medical oxygen—and this was before the pandemic.

Bernard Olayo, a health specialist at the World Bank, is trying to do something about it. After graduating from medical school in the mid-2000s, he worked at a rural hospital in his native Kenya, where many of the patients were children with pneumonia who needed oxygen for treatment. But there was never enough oxygen available. Often, several patients had to share a single cylinder of oxygen. When there wasn't enough to go around, Olayo and his colleagues would have to decide who would get it and who would go without—a gut-wrenching choice that often meant one child would live and another would die.

Olayo set out to learn why something as seemingly basic as oxygen was so hard to come by in Kenya. One problem, he learned, was that there was only one oxygen supplier for the entire country, and because there was no competition, the supplier could charge exorbitant prices. (At the time, oxygen in Kenya cost about thirteen times more than in the United States.) Plus, many Kenyan health facilities are hundreds of miles from the nearest oxygen plant, which caused two other problems: Transportation costs added to the price, and poor roads added to the delivery time. New supplies were often delayed and sometimes never arrived at all.

In 2014, Olayo created an organization called Hewatele—the Swahili word for "abundant air"—to try a different approach. With funding from local and international investors, Hewatele built oxygen plants at several of the busiest hospitals in the country, where demand is highest and reliable electricity for production is available. It devised a milkman model: Oxygen cylinders would regularly be dropped off at remote hospitals and clinics and the empty cylinders returned for a refill. Using this new approach, Hewatele has cut the market price for oxygen in Kenya by 50 percent and reached some 35,000 patients. And as I write this, the group is looking to expand within Kenya and to other parts of Africa.

In addition to needing oxygen, patients who are severely sick might need to be intubated—have a tube put down their windpipe—and use a ventilator to help them breathe. In extreme cases, people's lungs might be so badly damaged that they can no longer oxygenate the blood, and a machine will need to do it for them. Just as medical oxygen itself was already hard to come by in many lower-income countries before COVID, so were the medical expertise and equipment needed to administer it. The pandemic has made the problem many times worse.

A recurring theme of this book is that we don't have to choose between preventing pandemics and improving global health more broadly—they reinforce each other. This is a classic example: If we do a better job of equipping the world's health systems with oxygen and other tools, as Hewatele is doing, then more health care workers will have the equipment they need to deal with everyday problems like pneumonia and premature births. And during a crisis, like an outbreak that's threatening to become a pandemic, they'll be able use this equipment and their expertise to save lives and stop the disease from overwhelming the entire health system. Each makes the other stronger.

Treating disease is nothing new to humans. The practice of using roots, herbs, and other natural ingredients as healing agents dates to ancient times. Some 9,000 years ago, Stone Age dentists in modern-day Pakistan drilled into their patients' teeth with pieces of flint. The ancient Egyptian physician and scientist Imhotep cataloged treatments for 200 diseases around 5,000 years ago, and the Greek physician Hippocrates prescribed a form of aspirin—extracted from the bark of the willow tree—more than 2,000 years ago.

But it's only in the past couple of centuries that we've been able to synthesize medicines in the lab rather than by extracting them from things we found in nature. One of the earliest synthesized drugs was created in the 1830s, when several scientists and physicians working independently all managed to make chloroform, the powerful anesthetic and sedative that, among other uses, would help Queen Victoria through the pain of childbirth.

Sometimes a drug has been invented because some enterprising scientist set out to do it, but sometimes it happens by pure accident, as in 1886, when two young chemistry students at the University of Strasbourg stumbled on a solution to a problem they weren't even looking to solve. Their professor was investigating whether a substance called naphthalene—a by-product of making tar—could be used to cure intestinal worms in humans. They administered the naphthalene with surprising results: It didn't get rid of the worms, but it did break the person's fever. After further investigation to figure out what had happened, they realized that they hadn't used naphthalene at all, but rather a then-obscure drug called acetanilide, which the pharmacist had handed them by mistake.

Soon, acetanilide was on the market as a cure for fevers and as a pain reliever, but doctors found that it had an unfortunate side effect: It made some patients' skin turn blue. After additional research, they learned that a substance could be derived from acetanilide that would have all the benefits without the blue hue. It was called paracetamol, which Americans know as acetaminophen, an active ingredient in Tylenol, Robitussin, Excedrin, and a dozen other products that you may have in your medicine cabinet right now.

Even in modern times, drug discovery still relies on a mixture of good science and good luck. Unfortunately, when an outbreak appears to be headed toward a pandemic, there's no time to count on luck. We'll need to develop and test treatments as fast as possible, much faster than we did for COVID.

So let's suppose we're in that situation: There's a new virus that looks like it could go global, and we need a treatment. How will scientists go about making an antiviral?

The first step is to map the virus's genetic code and then, armed with that information, figure out which proteins are the most important in the life cycle of the virus. These essential proteins are known as the targets, and the search for a treatment essentially boils down to defeating the virus by finding things that will keep the targets from working the way they should.

Until the 1980s, researchers trying to identify promising compounds had to get by with only a rudimentary understanding of the targets they were seeking. They would make their best educated guess and run an experiment to see if they were right; most of the time, they weren't, and they would move on to the next molecule. But the tools available for identifying the right drug have gotten much better in the past forty years, with the advent of a field called structure-guided discovery.

In structure-guided discovery, instead of testing each possible compound in a lab, scientists can program computers to create 3D models of parts of the virus that help it function and grow, and then design molecules that attack those targets. Moving the search for compounds from laboratory experiments to structure-guided discovery is like playing chess on a computer instead of a board—the game still happens, but not in a physical space. And just as with chess, structure-guided discovery has become more sophisticated with the growing processing power of computers and advances in artificial intelligence.

Here's how it worked for Paxlovid, the antiviral announced by Pfizer in late 2021. Scientists had identified how COVID hijacks parts of your cells to make more copies of itself (these parts are sequences of amino acids, the building blocks of proteins). Using this knowledge, they designed a molecule that operates like an undercover cop conducting a sting operation. It mimics most of the sequence of amino acids that COVID will seek out, but it's missing key pieces of the sequence, so it disrupts the life cycle of the virus. There are several stages of the life cycle that can be disrupted. In the case of HIV antivirals, by far the biggest category of antivirals, we have ones that attack each stage, and we combine three of them so that the virus is very unlikely to mutate to stop them from working all at once.

Even though scientists can now run virtual experiments very quickly on a computer, sometimes they still need to do the real thing—to match a compound with the protein from a virus in a lab and see what happens. But technology is changing this approach too.

In a process known as high-throughput screening, robotic machines can run hundreds of experiments at a time, mixing compounds and proteins and then using various methods to measure the reaction. With high-throughput screening, companies can now test millions of compounds in a matter of weeks, a task that would normally take a team of humans years to complete. Many of the major pharmaceutical companies have collected millions of compounds; if each collection is a library, then high-throughput screening is the fast, methodical way to search through every book on the shelf for just the right word.

And even if there isn't a good match—if there isn't an existing compound that looks as if it might make a good treatment—that's helpful information. The faster an existing compound can be ruled out, the faster scientists can move on to making new molecules.

Regardless of the method involved, once a promising compound is identified, the scientific teams will analyze it to determine whether it's worth further exploration. If it is, a different team—the medicinal chemists—will try to optimize the compound in a process that's a bit like squeezing a balloon. They might tweak it in one way to make it more potent, but then discover that the higher potency also makes it more toxic.

Once they've found a promising candidate in the exploratory phase, they will spend a year or two in the preclinical phase, studying whether the candidate is safe at effective doses and whether it actually triggers the expected response in animals. Finding the right animal is not as easy as it sounds, because they don't always respond to a drug the way that humans will. Researchers have a saying: "Rats lie, monkeys exaggerate, and ferrets are weasels."

If all goes well in the preclinical phase, we'll move into the riskiest and most expensive part of the process: clinical trials in humans.

In May 1747, a physician named James Lind was serving as a ship's surgeon on the British Royal Navy vessel *Salisbury.* He was horrified by the number of sailors who were suffering from scurvy, a condition that causes muscle weakness, exhaustion, bleeding from the skin, and eventually death. No one knew at the time what caused scurvy, but Lind wanted a cure, so he decided to try various options and compare the results.

He selected twelve patients on board who had similar symptoms. They all ate the same food—gruel sweetened with sugar in the morning, mutton broth or barley and raisins for dinner—but were given different treatments. Two drank a quart of cider each day. Two others were given vinegar. Other pairs of patients got either seawater (the poor souls), oranges and a lemon, a medicine concocted by a hospital surgeon, or a mixture of sulfuric acid and alcohol known as elixir of vitriol.

The citrus treatment won out. One of the two men who received it was back on duty in six days, and the other recovered quickly enough to begin tending to the rest of the patients. Although the British navy wouldn't make citrus a required part of a sailor's diet for nearly fifty years, Lind had found the first real evidence of a cure for scurvy. He had also run what is widely regarded as the first controlled clinical trial of the modern era.*

Other innovations in clinical trials would follow in the decades after Lind's experiment: the use of placebos in 1799, the first

* We now know that scurvy is caused by a deficiency of vitamin C. May 20, the day Lind began his trial, has been designated International Clinical Trials Day.

double-blind trial (in which neither the patient nor the physician knows who is getting what treatment) in 1943, and the first international guidance on the ethical treatment of trial participants in 1947, following revelations of the Nazis' horrific experiments during World War II.

In the United States, a series of laws and court rulings over the course of the twentieth century slowly built up the testing and quality assurance regime that exists today. It's this process that our hypothetical treatment for a new pathogen will have to go through. Let's track how it typically works, phase by phase.

Phase 1 trials. With permission from your government's regulator—in the U.S., it's the Food and Drug Administration—to run clinical trials in humans, you'll start with a small trial involving a few dozen healthy adult volunteers. You're looking to see whether the drug causes any adverse effects and to zero in on a dosage that's high enough to produce the effect you want but not so high that it makes the patient sick. (Some cancer drugs are tested only on volunteers who already have the disease because they're too toxic to be given to healthy people.)

Phase 2 trials. If everything goes well and you know your drug is safe, you're allowed to move on to larger trials. Here you'll give it to several hundred volunteers in the target population—people who are sick and otherwise fit the profile—and you'll look for proof that the drug does what you hope it will. Ideally, at the end of Phase 2, you know the drug works and you have the right dosage, because the next phase is so expensive that you only want to move ahead if you've got a good chance of success.

Phase 3 trials. If all goes well up to this point, you'll run even larger trials involving hundreds and sometimes thousands of sick volunteers—half of whom get your drug candidate, the other half of whom get either the standard treatment for their condition or, if there isn't a treatment yet, a placebo. This is a lot smaller than Phase 3 trials for vaccines, which I'll explain in the next chapter.

Everyone is already sick with the disease you're trying to treat, so you can see a lot faster whether the drug is working. (If there's already a treatment on the market, you'll have to enroll more volunteers, because you'll need to show that your product is at least as effective as the competition.)

Another hurdle in Phase 3 is finding enough volunteers to make sure your drug candidate is safe and effective for everyone who might take it. You need to find people who are sick—obviously at this stage there's no point in giving your potential cure to people who don't have the disease—but for the reasons we covered in Chapter 3, it's hard enough to identify those people, much less to identify the ones who are not only sick but also willing to volunteer to try out a new drug. And because everything from age to race to overall health can affect how a drug works in a person's body, it's important to study how a lot of different people react to it. Recruiting a diverse set of patients for a clinical trial can sometimes take more time than running the trial itself.

Regulatory approval. If you get out of Phase 3 and believe your drug is safe and effective, you go back to the regulatory agency and apply for approval. The application typically includes hundreds of thousands of pages, and in the U.S., the FDA's review can take a year or—if there are concerns with the application—even longer. The agency will also inspect the factory where you're going to make the drug, and they'll review the label you want to put on the bottle as well as the printed information that's included in the packaging. Even after you're licensed, you might be required to run another phase of trials among certain groups of people, and in any case, regulators will continue to check your production line to make sure the doses you're making are safe, pure, and potent. And as more people take your drug, you'll keep watching out for adverse effects (an especially rare problem might only show up once lots of people are taking it), and you'll also be on the lookout for signs that the pathogen is building up a resistance to your drug.

Now, this is how it works in nonpandemic times. In the emergency that COVID was, it needed to happen much, much faster. The U.S. government and other funders put up the money for some of the Phase 3 trials—the most expensive step in the process, because it involves so many people—even before drug candidates had gone through Phase 1. Scientists also put off studying aspects of the drugs that weren't crucial in an emergency while maintaining the key safety aspects. It was akin to proving that a car will get you where you need to go without exploding halfway there, but being a little unsure about its gas mileage or how well the tires will handle snow.

In the early days of COVID drug trials, there were few standard protocols for the trials or agreements about what data to collect even within countries. That led to a lot of wasted time and effort, as multiple poorly designed clinical trials tested the same products but didn't produce conclusive evidence. Often, by the time the protocol was written and approved for a trial in a given place, the number of cases in that place had dropped so low that the trial couldn't be run effectively anymore. We need to standardize the approach to trials ahead of time, ensuring that they're well designed, are run in several places, and are set up to provide definitive evidence as fast as possible. One of the few trials that was handled well was the RECOVERY trial in the U.K., which looked at a number of drugs, including dexamethasone: It was ready to go within six weeks and included 40,000 participants at 185 sites.

The RECOVERY trial was one of many efforts supported by a new project called the COVID-19 Therapeutics Accelerator,* which was designed to speed up the process of finding COVID treatments and then make sure that millions of doses were available for people in low- and middle-income countries. It helped coordinate drug trials and, to make it easier to identify people who might be eligible

* Initially launched by the Wellcome Trust, Mastercard, and the Gates Foundation.

to participate in these trials, the Accelerator also helped develop new diagnostic tools. As of the end of 2021, donors had committed more than $350 million to this effort.

Some new ideas may push the limits of what regulators are comfortable with. One is that, if you test positive, you would get an immediate text message giving you the chance to volunteer for a clinical trial that needs people with profiles like yours. Simply click "Enroll Me" and you'll start the process; if you're selected, you'll get access to treatment—either the one being studied or the best one already in use—and you'll be helping speed up the clinical trial. Another innovation I hope to see is putting regulatory submissions in the cloud and in a standard format, so that they can be reviewed by all the regulatory agencies across the globe without duplication. And in the United States especially, adopting a standard format for patients' health records would have many benefits, including making it easier to find potential volunteers for drug trials.

There are even more ways to simplify and shorten the process of testing new treatments—including a controversial approach known as the human challenge study. These trials are already being carried out for malaria drugs: Volunteers agree to allow themselves to be infected with the malaria parasite so that researchers can test the potential impact of new drugs, antibodies, and vaccines. The reason it's ethical to do this is because it is done with healthy adults who are treated with effective antimalarials as soon as they start feeling sick. These human challenge studies have dramatically accelerated work on malaria treatments and vaccines, because you don't have to wait for people to get the disease naturally before you can start learning whether a new product works.

There's a similar option for a viral infection like COVID, when the risks to healthy young adults are minimal and we have effective treatments that can be given to volunteers once they start showing symptoms. If we can overcome the scientific challenges and work through the ethical issues, carefully conducted human challenges could replace many of the complicated studies that require finding high-risk patients early in the course of the disease—giving researchers a quick early read on the promise of new potential therapies.

So let's return to our hypothetical example of a new pathogen. We've now developed a treatment, run trials to prove that it's safe and effective, and gotten the green light to make and sell it. It's time to start manufacturing. Although making a small-molecule drug is easier than making an antibody, which in turn is generally easier than making a vaccine for reasons I'll explain in the next chapter, it's still worth taking a moment to walk through the challenge of scaling up manufacturing.

First, a team of chemists will work on finding a consistent way to produce the key part of our drug—known as the active pharmaceutical ingredient—by setting off a series of reactions using chemicals and enzymes. The best path might involve as many as ten separate steps: The chemistry team will start with certain ingredients, spark a reaction among them, capture the by-products, use some of those in another reaction, and so on, until they have the active ingredient we're after. Then they'll turn it into the form in which patients will take it—such as a pill, nasal spray, or injection.

Quality control is relatively easy for small-molecule medications, as opposed to vaccines. Because the product is just a string of molecules and not a living thing, we can use analytical tools to confirm that it has all the necessary atoms in all the right places.

This one fact is a godsend for everyone who cares about equity in global health, because it gave rise to one of the most important innovations in this field in the past several decades: generic drug manufacturers committed to creating high-quality, low-cost versions of lifesaving drugs.

Historically, the companies that invent new drugs have been based in higher-income countries. Because it costs so much to develop a new product, they try to recoup their costs as quickly as possible by selling doses at the higher prices that rich countries can afford. It doesn't make sense to tinker with the manufacturing process in order to lower the cost of making the product (by reducing the number of steps involved, for example), because that would require going through some of the regulatory process again, and even then, you'd save only a small fraction of the overall cost of production. This can mean that the cost stays too high for developing countries, and it's why it sometimes takes decades for drugs that are widely available in the rich countries to reach poor ones.

This is where low-cost generic manufacturers come in. Part of their role is to help people in poor countries get access to the same drugs and other lifesaving inventions that are widely available in rich countries.*

Generics made their mark on global health around two decades ago. At the time, lifesaving HIV drugs were too expensive for countries like Brazil and South Africa, which meant that millions of people living with HIV were priced out of the market. So generic manufacturers began to duplicate the drugs in violation of the intellectual property rights of the companies that had invented them, and the governments in those countries did little to enforce the patents on the original drug. At first, the patent holders objected, but they eventually backed off after realizing that a tiered pricing approach would work better. They made information about their drugs available to low-cost generic manufacturers, which were allowed to sell to developing countries without paying any royalties. In this tiered pricing approach, the highest price is charged in rich countries, a lower price in middle-income countries, and the lowest possible price—one that is only marginally more than the cost of manufacturing—in low-income countries.

* Generic manufacturers are also the reason why you might be able to get significantly cheaper versions of some of the prescriptions you take.

One problem is that once a drug is made generic, there are few incentives to invest in reducing the manufacturing cost, since other companies could immediately copy their improvements. To resolve this, donors will hire experts and fund the optimization work and the upfront costs of implementing a new process. In 2017, for example, the Gates Foundation and a number of partners helped create a generic form of a more effective version of an HIV drug cocktail, work that was enabled by a free license from the pharmaceutical companies that invented the drugs.

The generic firms were able to reduce the cost so much that today nearly 80 percent of people who get HIV treatment in low- or middle-income countries are receiving the improved cocktail. The new drug requires a much lower dose—and a smaller pill—to be effective than previous treatments did, which means it's a lot easier for people to take. It also causes fewer side effects and is less likely to cause drug resistance.

Of course, the generic-drug business has downsides. As they headed for low prices and their profit margins narrowed, a few generic producers haven't maintained the quality of their products the way they should have. But those are the outliers, and it's hard to overstate the positive impact of low-cost, high-quality, high-volume generic manufacturers. Months before studies proved that molnupiravir is an effective antiviral, Merck had already negotiated licensing deals with several generic manufacturers in India, allowing them to make and sell generic versions in India and more than 100 other low- and middle-income countries. Researchers developed ways to drive down the cost of making it, and other organizations helped the generic companies get ready to make the drug and apply for approval from the WHO. In January 2022—just two months after molnupiravir's successful results had been announced—generic companies made 11 million doses available to low- and middle-income countries, a first step to producing many more.

Generic manufacturers produce the large majority of medicines used by people in low- and middle-income countries.* The WHO's malaria program, which works largely with generic manufacturers, estimated that it will eventually help 200 million people get drugs to treat malaria who wouldn't have gotten them otherwise. Even in the United States, 90 percent of all prescriptions filled are for generic drugs.

I wish that making antibodies were as straightforward as making drugs. To produce antibodies for the hypothetical pathogen we're trying to contain, we'll need to find patients who've survived the disease, draw their blood, and identify the antibodies that their bodies developed to fight this particular infection. Since their blood will contain antibodies for essentially every disease they've ever encountered, we'll have to isolate the one we're looking for by introducing the virus to a bit of their blood, and then watching to see which antibodies stick to it. Those are the ones we want. (An alternative is to go through the same process but with blood from humanized mice—rodents in which human cells or tissues have been implanted.)

Once we've isolated the right antibody, we'll need to copy it billions of times over. We'll likely do that by growing them on the CHO cell platform, which consists, as you surely guessed, of ovarian cells from Chinese hamsters.

These cells are so useful because they're especially hardy, they can be maintained indefinitely, and they grow quickly. Most of the ones in use around the world today are clones of a cell line created by a geneticist named Theodore Puck, who worked at the University of

* A few examples: Dr. Reddy's Laboratories, Aurobindo, Cipla, and Sun (all based in India), Teva (based in Israel), and Mylan, which is now part of Viatris and Sandoz (in the U.S. and Europe).

Colorado Medical School in 1957. He had managed to get his hands on a single female hamster whose ancestors had been smuggled out of China in 1948, just as the Communists were ousting the Nationalists in the Chinese civil war.

Unfortunately, the CHO platform doesn't produce antibodies fast enough to meet much of the need during a pandemic. The world produces 5 to 6 billion doses of vaccines every year, and only about 30 million doses of antibodies. CHO antibodies are also expensive to make—the current cost of producing them is in the range of $70 to $120 per patient, too high for many low- and middle-income countries. But scientists are working on ways to solve these problems.

For example, some are looking into different host cells that would produce antibodies more efficiently. Others are studying ways to find more potent and highly selective antibodies, so that you wouldn't need as much product per patient. Already there are ideas being tested but not yet commercialized that would get the cost down to more like $30 or $40 per dose. But the ideal would be to cut costs by a factor of ten—getting them below $10 per person—while also producing ten times as many doses in the same amount of time. It will take a number of improvements to reach that goal, but once we have these promising tools, they will be able to help more people around the world.

Companies are also developing solutions to the problem of variants. One approach involves making antibodies that target parts of the virus that don't change, even across different variants—meaning they're just as effective against the variants as they are against the original virus. Another approach involves mixing a cocktail of antibodies that attack different parts of the virus, making it much harder for the virus to develop resistance to them.

—

Back to the hypothetical disease that we're making treatments for. Let's assume we get a treatment approved and are able to manufacture lots of doses. How do we make sure it actually reaches everyone who needs it?

Even if the cost is low, some countries will need donations to be able to get enough for all their people. For decades, low- and middle-income countries have been getting help from various organizations to buy and deliver medicines. You've probably heard about the very effective UNICEF; one that's not as well known is the Global Fund, which helps countries buy medicines and other tools to fight HIV, TB, and malaria. The Global Fund is now the world's largest financer of these efforts, reaching more than 100 countries, and in 2020 it expanded its purview to include COVID supplies.

Of course, cost isn't the only hurdle we'll need to overcome. Even once we have a cheap treatment, it may be hard to get it to the patients who need it. And we'll need to make sure they get the right treatment at the right time. (Remember, for example, that mAbs and antivirals need to be given shortly after symptoms start, while a steroid like dexamethasone is only appropriate later in the disease, when the patient is severely ill.)

Even then, something as seemingly basic as the packaging of the medicine might make people reluctant to take it. Some HIV drugs also help prevent people from getting infected in the first place—an approach called pre-exposure prophylaxis—but many patients don't want to take an AIDS medicine out of fear that others will think they're HIV-positive. That problem can be solved, but not without some effort, because you can't just start making pills that look different. You have to test every element, including the shape, size, and even the color of the pill.

There are still more barriers to reaching people in low-income countries. Before a company launches a new drug in a market where it expects to make a big profit, it spends years figuring out how to target the right patients and training health care workers to use the new drug.* In fact, it might spend as much money doing this work as it did developing and making the drug itself! But when most of the people who need a drug live in poor countries, companies tend to spend very little time or money laying this groundwork. And the situation is even worse during a major outbreak or pandemic, because there's little or no time to communicate with providers and patients early on—so it's not surprising if people don't immediately take up new medicines, or if they're confused about how to use them.

I feel sure we will have better treatment options in the next big outbreak than we did for COVID. One of the keys to making that happen will be big libraries of drug compounds that we can quickly scan to see if existing therapies work against new pathogens. We have some of these libraries already, but we need more. This will require substantial investment to bring together academia, industry, and the latest software tools.

We need libraries that cover many types of drugs, but there are some types that should be a top priority. The most promising, in my view, are the ones known as pan-family and broad-spectrum therapies—either antibodies or drugs that can treat a wide range of viral infections, especially those that are likely to cause a pandemic. We could also find better ways of activating what's known as innate immunity, which is the part of your immune system that kicks in just minutes or hours after it detects any foreign invader—it's your body's first line of defense. (It stands in contrast to the adaptive immune response, which is the part that remembers the pathogens you've encountered before and knows how to fend them off.) By boosting your innate immune response, a drug could help your body stop an infection before it takes hold.

* Sometimes they go too far, as some companies did with opioids.

To deliver on these promising approaches, the world needs to invest more into understanding how various dangerous pathogens interact with our cells. Scientists are working on ways to mimic these interactions so they can quickly figure out which drugs might work in an outbreak. A few years ago, I saw a demonstration of a "lung on a chip," an experimental device you could hold in your hand that operated just like a lung, allowing researchers to study how different drugs, pathogens, and human cells affect one another.

With advances in artificial intelligence and machine learning, it's now possible to use computers to identify weak spots on pathogens that we already know about, and we'll be able to do the same when new pathogens arise. These technologies are also speeding up the search for new compounds that will attack those weak spots. With adequate funding, different groups could take the most promising new compounds through Phase 1 studies even before there's an epidemic, or at least have several leads that can be turned into a product quickly once we know what the target looks like.

In short, although therapeutics didn't rescue us from COVID, they hold a lot of promise for saving lives and preventing future outbreaks from crippling health systems. But to make the most of that promise, the world needs to invest in the research and systems we'll need to find treatments much faster, and to deliver them to people who need them, wherever they are. If we succeed in doing this, the next time the world faces an outbreak, we'll minimize the disruption and save millions of lives.

CHAPTER 6

GET READY TO MAKE VACCINES

It is easy to forget, now that billions of people have received at least one dose of a COVID vaccine, that the odds were not in humanity's favor. I mean, they were *really* not in our favor.

The fact that scientists were able to create multiple successful COVID vaccines is itself unusual in the history of disease. The fact that they did it in roughly a year is miraculous.

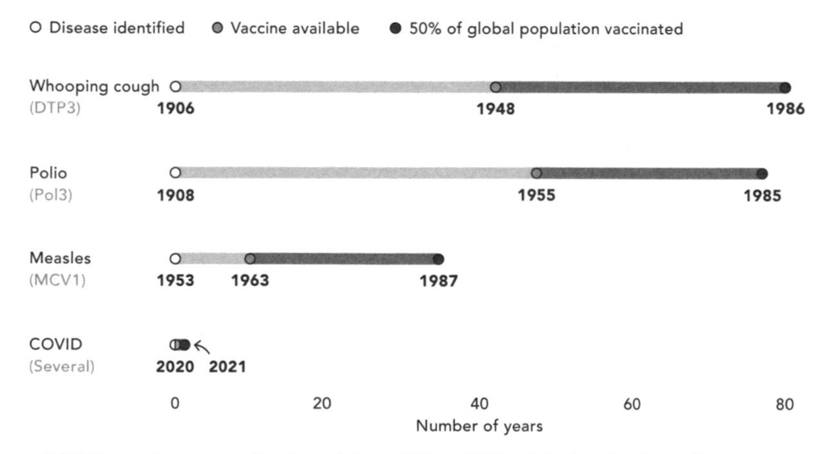

COVID vaccines were developed incredibly quickly. It took scientists only one year to create safe and effective vaccines for the virus. In comparison, eighty years passed between identifying whooping cough and immunizing 50 percent of the world's population against it. (Our World in Data)

Pharmaceutical companies, being the data hounds that they are, have a way of measuring the chance that a drug or vaccine candidate will make it all the way through the arduous process of being approved for use in humans. This measure, which is called the Probability of Technical and Regulatory Success, depends on several factors, including whether similar products have proved successful. If you're testing a vaccine that works more or less the same way as one that has already been approved, your odds are better.

Historically, the average probability of success for vaccine candidates is 6 percent.* What this means is that if you start with 100 candidates, only 6 of them will make it all the way to full regulatory approval. The others will fail for any number of reasons—they might not give people enough immunity, or clinical trials might not give you the conclusive results you need, or they might have unintended side effects.

Of course, this 6 percent figure is just an average. The odds for drugs and vaccines created using tried-and-true methods would be a few percentage points higher. And they're a few points lower when you're trying a new approach. First you have to prove that the basic approach works. Then you may also have to prove that the specific vaccine you make using that approach works. You have to run massive trials, which can involve hundreds of thousands of people, and later you'll have to watch for side effects in millions of people. There are hurdles everywhere.

Fortunately, COVID is relatively easy to target with a vaccine, partly because the spike on its surface is not as camouflaged as the proteins on some other viruses. That's why the success rate for COVID vaccines has been unusually high.

* A vaccine candidate is just what you'd imagine: something that might turn out to be a safe, effective vaccine, but is still in the development stages. It's like a bill that is making its way through Congress or Parliament and may or may not become law.

Yet the greatly underappreciated miracle of COVID vaccines is not the fact that they were created and approved. It's that they were created and approved faster than any other vaccines ever.

In fact, it happened even more rapidly than many people, including me, were willing to predict in public. In April 2020, although I thought we could have a vaccine by the end of the year, I wrote on my blog that it could take as long as twenty-four months—I thought it wouldn't be responsible to raise the prospect of quick successes if there was a decent chance they wouldn't happen. In June, after seeing initial data on some promising vaccine candidates, a former commissioner of the Food and Drug Administration told *The New York Times,* "Realistically, the 12 to 18 months that most people have been saying would be a pretty good marker but still optimistic."

What did in fact happen was the best-case scenario. The vaccine made by Pfizer and BioNTech was approved for emergency use at the end of December, just a year after the first COVID cases were identified.

To get a sense of just how fast this happened, consider that the vaccine development process—from the first discovery in a lab, through proving that it works and getting it licensed—typically takes between six and twenty years. It can take as many as nine years simply to get a product ready for clinical testing in humans, and even with a lot of time, there's no guarantee of success. The first trial for an HIV vaccine began in 1987, and we still don't have one that's licensed.

Before COVID, the land speed record for developing a vaccine was four years. This remarkable feat was achieved with a vaccine for mumps by a scientist named Maurice Hilleman, who was one of the most productive vaccine makers ever. Of the fourteen vaccines currently recommended for children in the U.S., he and his team at Merck Pharmaceutical developed eight, including ones that protect you from measles, hepatitis A and B, and chickenpox.

In 1963, his five-year-old daughter, Jeryl Lynn, came down with a

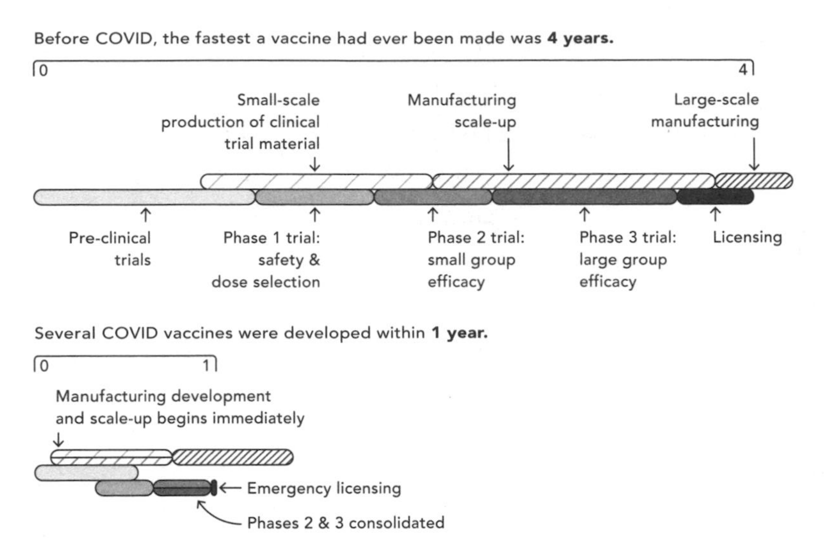

Making a vaccine. All vaccines go through a rigorous process to make sure they're safe and effective. Several vaccines were created for COVID within a year by consolidating development stages without sacrificing safety. (NEJM)

sore throat. Suspecting that she had mumps—for which there wasn't yet a licensed vaccine—he took a sample from her throat using a swab, went to his lab, and isolated the virus. Eventually he used it to create the first licensed mumps vaccine in 1967. This strain of mumps is still used to make vaccines today, and it's named for his daughter. If you've received the MMR vaccine—for measles, mumps, and rubella—you got the Jeryl Lynn strain.

In Hilleman's time, making a vaccine in four years was a fantastic accomplishment. But one of the reasons he was able to move relatively fast is that there weren't the same stringent ethical standards for gaining consent or assuring quality that we have today. In any case, when an outbreak is threatening to become a pandemic, four years would be a disaster.

The implications for preventing pandemics are clear: We need to raise the odds of success for vaccines, and we need to reduce the time it takes to get them out of the lab and into humans, without sacrificing safety or effectiveness. We also need to manufacture so many of them so quickly that they'll be available to everyone in the world six months after the pathogen has been identified.

That's an ambitious goal, and as I mentioned in the Introduction, one that will strike some people as outlandish. But I'm convinced that it's possible, and in the rest of this chapter, I want to make the case that it's well within reach.

Getting a vaccine out of the lab and into recipients takes four steps—developing it, getting it approved, manufacturing it at high volumes, and delivering it—and we'll look at the opportunities to accelerate the process along the way. We'll look at why it can be so hard to create and test a vaccine and why it takes so long. What exactly is happening in those five or ten years before it's ready for the market? We'll also explore why scientists were able to move so much faster this time—it's a fascinating story of farsighted planning, dogged research by two heroic scientists, and more than a little luck.

Unfortunately, as we've seen with COVID, it's one thing to get a vaccine created and approved. It's another challenge entirely to avoid creating a system of vaccine haves and have-nots—to make and distribute enough doses to quickly reach everyone who needs them, including people in low-income countries who have a high risk of getting severely sick.

The distribution of COVID vaccines in 2020 and 2021 was, to

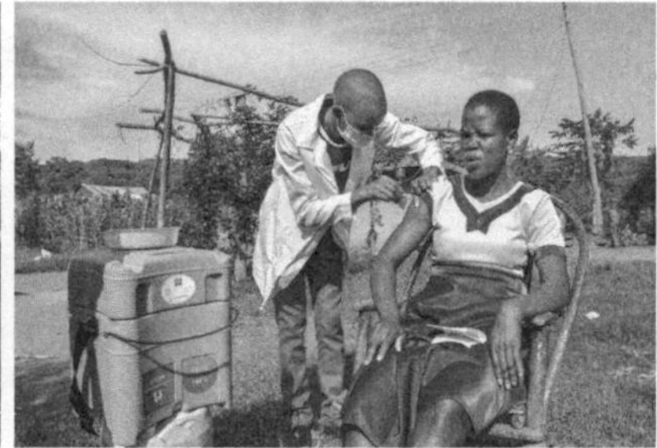

In the U.S., people lined up in cars at predetermined locations to get their vaccines, while many people in rural parts of low- and middle-income countries had to wait for limited doses to be delivered on foot.

quote Hans Rosling again, both bad and better. Vaccines reached more people faster than any other vaccination effort ever. They also reached many people in poor countries faster than ever—but not fast enough. So we'll look at ways to distribute vaccines more fairly.

Finally, we'll wrap up this chapter by discussing a new kind of drug that would complement vaccines—one that you could inhale and would keep the virus from getting into your body in the first place. Protecting yourself and others would be no more complicated than treating your hay fever.

My initiation into the world of vaccines came in the late 1990s, as I was learning about global health. When I discovered that children in poor countries were dying of diseases that kids in rich countries never died of—and that the main reason was that one group got certain vaccines and the other group didn't—I read up on the economics of immunization. It was a classic case of a market failure: Billions of people needed the great inventions of modern medicine, but because they had so little money, they had no way of expressing their needs in ways that matter to markets. So they went without.

One of our first major projects at the Gates Foundation was to help create and organize Gavi, the Vaccine Alliance,* an organization that pools donations to help poor countries buy vaccines. Gavi created a market where there wasn't one: Since 2000, it has helped vaccinate 888 million children and prevented some 15 million deaths. It's safe to say that Gavi is one of the foundation's contributions that I am most proud of, and in Chapter 8, I'll have more to say about how it works and the role it should play in preventing pandemics.

The more we learned about vaccines, the more I came to

* Formerly known as the Global Alliance for Vaccines and Immunization, or GAVI, it changed its name a few years ago to Gavi, the Vaccine Alliance.

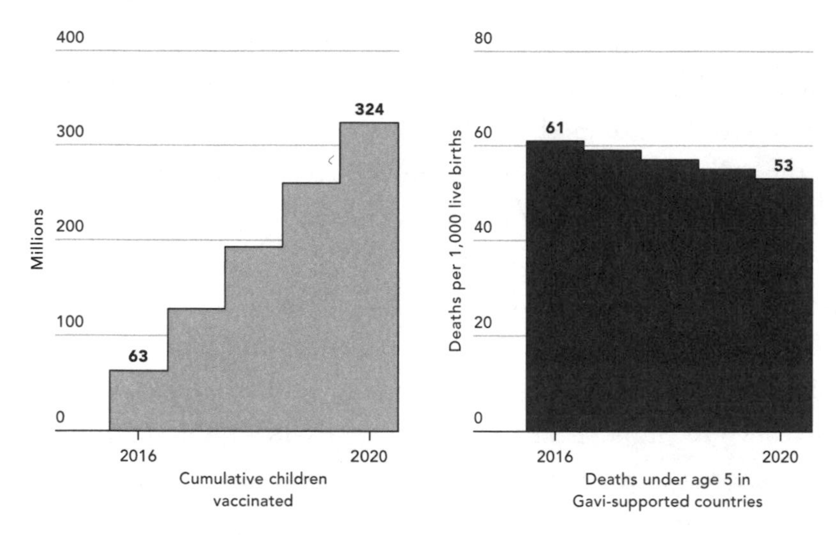

Gavi saves lives. Gavi has helped vaccinate 324 million children in the past five years alone. This chart shows how child mortality rates go down as vaccination rates go up. (Gavi, UN IGME)

understand about the science as well as the economics. It wasn't just that poor countries couldn't afford existing vaccines—they also didn't have the market power to demand new vaccines for diseases that primarily affected them. So the foundation started hiring people who were experts in making vaccines (and drugs). I had to learn a lot more about chemistry, biology, and immunology: I've spent countless hours in conversations with scientists and researchers from all over the world, and I've toured a lot of vaccine factories.

In short, I've spent a lot of time learning about the finances and operations of the vaccine industry, and I can say confidently that it is quite complex.

That's partly because we've decided as a society that we'll tolerate very little risk when it comes to vaccines. This cautious approach makes sense: After all, you give vaccines to healthy people, and giving them a vaccine that has bad side effects defeats the purpose. (People won't take a vaccine if it's likely to have severe side effects.) As a result, the industry is highly regulated, and vaccines go through a lengthy, rigorous process of being tested and monitored. I'll explain shortly how it differs from the drug-approval process, but here's just one example of how rigorous the process is: If you build a vaccine factory, you have to meet standards that touch on nearly every aspect of the building, from the temperature of the air and the volume of airflow to the curvature of the wall corners.

Another reason the industry is so challenging is the nature of the product. Vaccines are made up of massive molecules, on the order of a million times heavier than the molecules that make up aspirin. Many are created in living cells—some flu vaccines, for instance, are typically grown in chicken eggs—and since living things are inherently unpredictable, you don't necessarily get exactly the same result each time. Yet getting almost exactly the same result each time is crucial to making a vaccine that's safe and effective. You need highly specialized equipment, and you need trained technicians to run it, and every time you make a new batch, there are half a dozen or more variables that could change the final product in subtle but important ways.

Once you've found a method of making a vaccine and developed one that's safe for humans, you have to reproduce this method every time you make it—that's the only way a regulator can confirm that you've gotten the same result you got previously. In contrast with checking a small molecule—where someone can examine it and say, "I don't particularly care how you made this thing; I can tell it has the right atoms in the right places"—checking a vaccine requires the regulator to watch how you make it and then, on an ongoing basis, to make sure you don't change anything. In fact, the dozens of complex experiments that a company has to develop in order to ensure consistency contribute significantly to the ultimate cost of a dose. Unfortunately, several promising COVID vaccines have been seriously delayed because of such matters—this is not an area where it's okay to cut corners. By comparison, reproducing something like software is a cinch. Once you've debugged your code, you can copy it as many times as you want without fear that something new will suddenly go wrong with it. If making copies of software had occasionally introduced new problems, the industry wouldn't have been nearly as successful.

It also costs a lot of money to develop a vaccine: Estimates of the total cost to get one developed and licensed range from $200 million to $500 million. The number goes even higher if you factor in the cost of all the failures you'll have along the way: One widely cited but debated study on drugs (not vaccines) put the total figure at $2.6 billion, and as I mentioned earlier, drugs are usually much less complicated to create than vaccines.

During an outbreak, vaccine companies also have to contend with high expectations from the public. People want a new vaccine that's safe and effective, they want it fast, and they want it to be cheap.

I'm not defending every decision that a pharmaceutical company has ever made about pricing a product, and I'm not asking anyone to feel sorry for the industry. But if we're going to tap into their expertise in developing, testing, and manufacturing drugs and vaccines—and there's no way to prevent or even stop pandemics unless we do—then we need to understand the challenges they face, the process they go through when they're deciding what products to work on, and the incentives that push those decisions in one direction or another.

You may have noticed that I keep using words like *business, industry,* and *markets,* with the implication that much of the work on vaccines is done by private companies. That's intentional. Although nonprofits, academic institutions, and governments have essential roles to play—funding basic research and deploying vaccines widely, among others—the private sector is almost always responsible for the final stages of developing vaccines and manufacturing them in mass quantities.

This has important implications for our efforts to keep future outbreaks from going global. Remember that our goal is to have no more pandemics, so although we definitely need to be able to produce enough vaccines for everyone on the planet in case a disease goes global before we can contain it, we would rather keep a disease from turning into a pandemic in the first place. So we'll need vaccines for regional outbreaks, in which potential recipients will number in the hundreds of thousands, rather than the millions or billions. This will change the pharmaceutical companies' incentives dramatically. If you run a drug company and are expected to turn a profit, why spend all the effort and funding required to develop a vaccine with only a small pool of potential buyers, especially if you have to set the price so low that you're unlikely ever to make money?

Simply relying on market forces won't do the trick. The world needs a plan for getting vaccine factories ready in advance and financing new vaccines. This plan should include money to prepare for vaccine trials and approvals, as the U.S. government did during COVID, putting up $20 billion to help move various vaccine candidates through the pipeline.

It should also include significant funding for research and development on vaccines and other tools, some of which should go to CEPI, the organization I mentioned in the Introduction that gives grants to academic centers and private companies to get them to develop vaccines and vaccine technologies. By the summer of 2021, CEPI had raised $1.8 billion for its COVID response, but donors have been less interested in funding work on future pandemics. That's understandable—when one disease is killing millions around the world, it's hard to get people to think about a disease that may emerge at some point in the future—but this funding is part of the billions we need to spend in order to save millions of lives and prevent trillions of dollars of economic losses in the future.

One area in which CEPI can contribute is the creation of vaccines that would be effective against entire families of viruses, also known as universal vaccines. Today's COVID vaccines teach your immune system to attack part of the spike protein on the surface of one specific coronavirus. But researchers are now working on vaccines that would target shapes that show up on *all* coronaviruses, including COVID and its cousins, and would even be likely to appear on ones that evolve in the future. With a universal coronavirus vaccine, your body would be ready to fight viruses that don't exist yet. Coronaviruses and influenza viruses should be the targets of these vaccines, since they have been responsible for the worst outbreaks of the past twenty years.

Finally, the world's vaccine plan should establish a way to allocate doses so they provide the greatest benefit for public health and don't simply go to the highest bidders. COVAX was intended to solve this problem during COVID, but for reasons that were mostly out of its control, it fell short of its goals. The idea was to pool the risk that's inherent in developing vaccines, with richer countries subsidizing lower-income ones. But rich countries essentially pulled out of that arrangement and instead negotiated their own deals with vaccine companies, putting COVAX at the back of the line and undermining its ability to negotiate with those same companies. In addition, it took longer than expected to approve two of the vaccines that COVAX was counting on, and for a time, COVAX wasn't allowed to export low-cost vaccines made in India to other countries. Despite all those challenges, COVAX has been the largest supplier of vaccines to the world's poorest countries. But the world will need to do better next time, a subject I'll return to in Chapter 9.

Of course, financing scientific work on new vaccines is only part of the equation. Those vaccines actually have to be developed—even faster than the COVID vaccines were—and by far, the most promising technology to do that is mRNA vaccines. For most people, these seemed to come out of the blue, but in fact they were the

Katalin Karikó, a Hungarian biochemist, helped develop the technology that is now used to make mRNA vaccines.

product of decades of painstaking work by researchers and product developers, including two who had to fight tooth and nail for their revolutionary idea.

From the time she was sixteen years old, Katalin Karikó knew that she wanted to become a scientist. She had a particular fascination with messenger RNA, or mRNA, molecules that (among other things) direct the creation of proteins in your body. In the 1980s, while working on her PhD in her native Hungary, she became convinced that tiny strands called messenger RNA could be injected into cells to allow the body to make its own medicines.

Messenger RNA functions as a kind of middleman—it carries the directions for making proteins from your DNA to the factories in your cells where the proteins will be assembled. It's a bit like the waiter in a restaurant who writes down your order and takes it to the kitchen, where the cooks will make your meal.

Using mRNA to make vaccines would be a major departure from the way most vaccines work. When you're infected by a virus, it invades certain cells in your body, uses the cells' machinery to make copies of itself, and then releases the newly made viruses into your blood. These new viruses go off in search of more cells to invade, and the process continues in a cycle.

Meanwhile, your immune system is primed to look for anything in your body with a shape that it hasn't seen before. When it encounters something it doesn't recognize, it says: "Hey, there's a new shape floating around. It's probably bad news. Let's get rid of it."

Ingeniously, your body is capable of going after both the viruses that are floating freely in your bloodstream and the cells that they have invaded. To defeat the ones in your blood, your immune system makes antibodies that latch on to that specific shape. (The cells that produce antibodies are known as B cells, and the ones that attack infected cells are called killer T cells.) Once you're producing antibodies and T cells, your body also produces memory B cells and memory T cells, which, as the name implies, help your immune system remember what the new shape looked like in case it shows up again.*

This system eventually stops the first attack of a virus and allows your body to do a better job the next time it sees that virus. But for viruses that make you sick—like COVID or influenza—it's better to prime your immune system so it can attack the virus the first time it shows up. That's what vaccines do.

Many conventional vaccines operate by injecting a weakened or dead form of thc virus you're trying to stop. Your immune system sees the new shapes on the virus, kicks into gear, and builds up immunity. With a weakened virus, there is always the question of whether it was weakened enough—if it wasn't, it might mutate back into a form that can cause disease. But if it was weakened too much, it won't activate a strong immune response in your body. Likewise, some killed viruses don't trigger much of an immune response. It takes years of lab work and clinical studies to make sure that conventional vaccines are safe and will produce a good immune response.

The idea behind mRNA vaccines was quite clever. Since mRNA takes the orders for proteins from the DNA and delivers them to the cooks in your cells' kitchen, what if we could change those orders in a very targeted way? By teaching your cells to make shapes that match shapes on the actual virus, the vaccine would trigger your immune system without having to introduce the virus itself.

If they could be made, mRNA vaccines would be a huge advance

* I'm simplifying things somewhat.

over conventional vaccines. Once you had mapped out all the proteins that make up the virus you wanted to target, you'd identify the one that you want antibodies to grab. Then you'd study the virus's genetic code to find the instructions for making that protein, and you'd put that code into the vaccine using mRNA. If, later, you wanted to attack a different protein, you'd just change the mRNA. This design process would take at most a few weeks. You would ask the waiter for fries instead of a side salad, and your immune system would do the rest.

There was just one problem: It was only a theory. No one had ever actually made an mRNA vaccine. What's more, most people in the field thought it was crazy to even try, not least because mRNA is inherently unstable and prone to degrading quickly. It was far from obvious that you could keep your modified mRNA together long enough to do its job. Also, cells have evolved to avoid being hijacked by foreign mRNA, and there would need to be a way of getting around this defense system.

In 1993, while doing research at the University of Pennsylvania, Karikó and her boss managed a feat that told them they were on to something: They got a human cell to produce a tiny amount of new proteins using a modified version of mRNA that had been cleverly altered so it could get past the cell's defense system.

This was a breakthrough, because it meant that if they could expand the production dramatically, they would be able to make a cancer treatment using mRNA. And although vaccines were not the focus of Karikó's work, other researchers saw that it would be possible to use mRNA to make those as well—for flu, coronaviruses, and maybe even various forms of cancer.

Unfortunately, Karikó's work lost momentum when her boss left academia for a biotech firm. She no longer had a lab or financial support for her work; although she applied for grant after grant, every application was rejected. The year 1995 was particularly dispiriting: She had a cancer scare, she was taken off the tenure track at work, and her husband was stuck in Hungary because of a problem with his visa.

But Karikó was undeterred. In 1997, she began working with Drew Weissman, a new colleague who came to the University of Pennsylvania with a promising background: He had done a fellowship at NIH under the supervision of Tony Fauci, and he was interested in using Karikó's work on mRNA to develop vaccines.

Together Karikó and Weissman kept pursuing the idea of working with mRNA that had been engineered in a lab. But they still had to get more mRNA past the cell's defense systems, a problem that other scientists helped solve.

In 1999, a cancer researcher named Pieter Cullis and his colleagues proposed that lipids—basically, tiny bits of fat—could be used to encase and protect a more delicate molecule, such as mRNA. Six years later, working with Cullis, biochemist Ian MacLachlan did it for the first time. The lipid nanoparticles he developed paved the way for the first mRNA vaccines.

As late as 2010, hardly anyone in the federal government or private industry was interested in trying to make vaccines using mRNA. Major pharmaceutical companies had tried and failed, and some scientists felt that mRNA would never trigger enough of a response in the body. But an official at DARPA, the little-known research program for the U.S. military, saw enough promise in the technology that he started funding mRNA vaccines for infectious diseases.*

As pioneering as this work was, it didn't lead immediately to new vaccines. Accomplishing that would be the task of companies dedicated to translating the breakthrough into a product that could be approved and sold; the U.S.-based Moderna and Germany-based CureVac and BioNTech were founded to do just that. In 2014, Karikó joined BioNTech, which was working on an mRNA vaccine for cancer.

* DARPA is the Defense Advanced Research Projects Agency.

Early efforts didn't work, although a test of a rabies vaccine showed promise. Still, Karikó and her BioNTech colleagues persevered, as did scientists at Moderna. When COVID hit, they immediately set out to make a vaccine for the new virus.

It was a good bet. The notion that mapping a virus's genome would allow you to create an mRNA vaccine in a matter of weeks proved to be exactly right.

In March 2020, just six weeks after scientists sequenced the COVID virus's genome, Moderna announced that it had identified an mRNA-based candidate and begun making it for clinical trials. On December 31, the mRNA vaccine made by BioNTech in partnership with Pfizer was approved for emergency use by the World Health Organization. When Karikó received the first dose of the vaccine she had done so much to create—a few days before it was officially approved—she wept.

It is hard to overstate the impact that mRNA vaccines have had on COVID. In many places, they account for virtually all of the COVID vaccinations. As of late 2021, more than 83 percent of vaccinated people in the European Union had received a vaccine made by either Pfizer or Moderna, both of which use mRNA. In the U.S., 96 percent had. Japan used *only* mRNA vaccines.

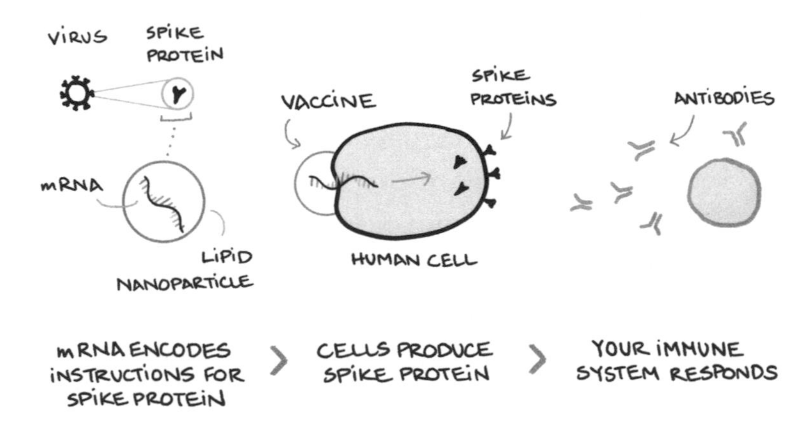

To me, the moral of the mRNA story is this: If the science makes sense, be willing to bet on crazy-sounding ideas, because they might be just the breakthrough you need. It took years of work to advance our understanding of mRNA enough that we could use it to develop vaccines. We are lucky COVID didn't come along five years earlier.

The agenda for mRNA researchers now is to keep making the technology better and broader—for instance, by going after vaccines for HIV and creating new ways to treat diseases. It may be possible to create a single mRNA vaccine that protects against several pathogens, rather than just one. And if we can find additional sources for the raw materials involved in making mRNA vaccines, their prices will go down.

In future outbreaks, we'll be measuring the time between the first case and the first vaccine candidate not in years or months, but in days or weeks. And mRNA will almost certainly be the technology that makes this possible.

If mRNA vaccines are the cool new kid on the block, viral-vectored vaccines are the equally cool kid who doesn't get quite as much attention because her family moved in a few years earlier.

Like mRNA, the viral-vectored approach was the subject of years of research and only recently produced vaccines that could be used in people. It works by delivering the spike or other target protein that you want your immune system to recognize as foreign. The delivery mechanism is a version of another virus—such as one that causes the common cold—that has been modified so it's harmless to humans; this virus, the carrier of the surface protein that the immune system will learn to make antibodies for, is what's known as the vector.

If you got a vaccine manufactured by Johnson & Johnson or Oxford and AstraZeneca, or if you got Serum Institute of India's Covishield, you got a viral-vectored dose. Although making the surface protein is harder than making the mRNA, these vaccines were still developed very quickly; the first two COVID vaccines that used viral vectors reached the market in just fourteen months, smashing the previous record for this approach. Before COVID, the only licensed viral-vectored vaccines were for Ebola, and those took five years to be approved.

Various types of COVID vaccines

Developers	Vaccine	Vaccine type	WHO emergency authorization date	Estimated doses shipped by end of 2021
Pfizer, BioNTech	COMIRNATY	mRNA	31 Dec 2020	**2.6 billion**
University of Oxford, AstraZeneca	VAXZEVRIA	Viral vector	15 Feb 2021	**940 million**
Serum Institute of India (Second source from Oxford/AstraZeneca)	Covishield	Viral vector	15 Feb 2021	**1.5 billion**
Johnson & Johnson, Janssen Pharmaceutical	J&J	Viral vector	12 Mar 2021	**260 million**
ModernaTX Inc, National Institute for Allergy and Infectious Diseases (NIAID)	SPIKEVAX	mRNA	30 Apr 2021	**800 million**
Sinopharm, Beijing Institute of Biological Products	Covilo	Inactivated virus	07 May 2021	**2.2 billion**
Sinovac Biotech Ltd.	CoronaVac	Inactivated virus	01 Jun 2021	**2.5 billion**
Bharat Biotech	COVAXIN	Inactivated virus	03 Nov 2021	**200 million**
Serum Institute of India (Second source from Novavax)	COVOVAX	Protein subunit	17 Dec 2021	**20 million**
Novavax	Nuvaxovid	Protein subunit	20 Dec 2021	**0**
Sanofi	Sanofi	mRNA	Development discontinued	**0**
University of Queensland, Commonwealth Serum Laboratories (CSL)	UQ/CSL (V451)	Protein subunit	Development discontinued	**0**
Merck, Institut Pasteur, Themis Bioscience, University of Pittsburgh	Merck (V591)	Viral vector	Development discontinued	**0**

There's another type of vaccine that has been around longer than either viral-vectored or mRNA vaccines. It's known as a protein subunit vaccine, and you may have received one to prevent flu, hepatitis B, or human papillomavirus infection (better known as HPV). Instead of using the whole virus to trigger your immune system, these vaccines introduce just a few key parts—hence the name "subunit." Because they don't use the whole virus, they're easier to produce than weakened or killed vaccines, but like both of those, protein subunits don't always elicit an immune response that's strong enough. So they may require what's known as an adjuvant, a substance that sets off red alerts in your immune system and shouts: "Hey, come look at this new shape over here! You'd better learn to attack it."

For COVID, the company Novavax created an adjuvanted protein subunit vaccine via a pretty complex process: They modified part of the gene that creates COVID's spike protein, inserted it into another type of virus, and then used that virus to infect cells taken from moths(!). The infected moth cells grew spikes like those on the coronavirus; once collected, these spikes were mixed with an adjuvant derived from the inner bark of the Chilean soapbark tree—which, believe it or not, is one of the most effective adjuvants in the world—and packaged into a vaccine. If you've encountered Nuvaxovid or COVOVAX, you've gotten a protein subunit vaccine.

As optimistic as I am about these technologies, I need to add a caveat: *We were good, but we were also lucky.* Because coronaviruses had already caused two previous outbreaks (SARS and MERS), scientists had learned quite a lot about the structure of the virus. Especially important was the fact that they had identified its characteristic spike protein—the tips on the crownlike virus you've seen a dozen pictures of—as a potential target for vaccines. When it came time to modify mRNA for a COVID vaccine, they had a sense of what the target should be.

The lesson here is that we need to pursue basic research on an even wider array of known viruses and other pathogens, so we understand as much as possible before the next outbreak. We should also be stepping up research on the broad-spectrum therapies that I mentioned in the previous chapter.

But no matter how quickly we make a new vaccine during an outbreak, it won't do any good if it takes years to get through the approval process. So let's look in detail at how that process works, and how we might accelerate it without sacrificing safety or effectiveness.

Humans invented vaccines long before they invented ways to make sure they worked. The British physician Edward Jenner is considered the founder of modern vaccines, having shown in the late eighteenth century that inoculating a young boy with cowpox—a disease related to smallpox, but with mild health effects—made him immune to smallpox too.* The word *vaccine* comes from the name of the cowpox virus, *vaccinia,* which in turn is derived from *vacca,* the Latin word for cow.

By the end of the nineteenth century, you could get immunized against smallpox, rabies, plague, cholera, and typhoid. But you couldn't be sure whether the vaccine was any good, or even whether it was safe.

This unregulated market had tragic consequences. In 1901, a contaminated smallpox vaccine caused an outbreak of tetanus in Camden, New Jersey. That same year, a contaminated serum that was supposed to offer protection from the bacterial infection diphtheria killed thirteen children in St. Louis.

Spurred by outrage over these incidents, the U.S. Congress began regulating the quality of vaccines and drugs, funding the Hygienic Laboratory of the U.S. Public Health Service in 1902. The job of regulation eventually moved over to the Food and Drug

* Like a lot of scientists at the time, Jenner had wide-ranging interests. He was an ornithologist and also studied the hibernation of hedgehogs.

Administration, and responsibility for federal research stayed with the Hygienic Laboratory, which we know today as the National Institutes of Health.

In the previous chapter, I walked through the process for getting a drug approved. It works largely the same way for vaccines, so I'll give just a quick summary here and note the major differences in getting approvals for each of them.

The exploratory stage. Two to four years of basic laboratory research designed to identify candidates.

Preclinical studies. A year or two assessing your candidate's safety and studying whether it actually triggers an immune response in animals.

Phase 1 trials. If you get permission from your government's regulator to run clinical trials in humans, you'll start a small one involving adult volunteers that looks very similar to the drug trials. There are some differences, though: Usually vaccine studies will have twenty to forty volunteers per cohort, in order to address the fact that different people have variable immune responses. At this point you're looking to see whether your vaccine causes any adverse effects, but to speed things up, companies might also try to combine Phase 1 and Phase 2 studies into a single protocol (J&J did this for its COVID vaccine). Small-molecule Phase 1 trials can be much smaller.

Phase 2 trials. You'll give the vaccine candidate to several hundred people who are representative of the population you want to reach, and you'll measure whether it's safe, check whether it bolsters their immune systems in the right way, and home in on the proper dosage.

Phase 3 trials. You'll run even larger trials involving thousands or tens of thousands of people, half of whom get a placebo or the most effective vaccine that's currently available. In Phase 3, you want to do two things, both of which require lots of volunteers from various communities where the disease you want to block is prevalent. One goal is to prove that the vaccine significantly reduces disease when compared with a placebo. Once the trial is under way, you have to wait until you have enough cases of sickness to know whether most of the infections occur among people who get the placebo rather than those who get the vaccine. The other goal in this phase is to identify relatively rare bad side effects, which might show up in, say, 1 in every 1,000 people who get the vaccine. So to have a chance of detecting 10 instances of the side effect, you'll need 20,000 volunteers: 10,000 to get the vaccine and 10,000 to get the placebo.

To make sure that your vaccine will work for everyone who needs it, you'll also require a diverse pool of volunteers from different genders, communities, races, ethnicities, and age groups. Stephaun Wallace, an epidemiologist at Fred Hutch in Seattle, is one of many people around the world who are trying to deepen the pool of potential volunteers.

Growing up Black in Los Angeles, he experienced firsthand how race shapes every part of how society treats people—including the medical system. After moving to Atlanta in his twenties, he created an organization that provided assistance to young Black men who were living with HIV. That experience ignited his interest in health inequities and led him to a career working to address them.

Wallace's work at Fred Hutch is particularly focused on improving the way clinical trials are run. He and his colleagues make a point of reaching out to diverse groups of people, including partnering with leaders from different communities, tailoring communications for those communities, making scheduling more flexible, and writing consent forms with accessible, nonscientific language.

Wallace was working on trials of potential HIV vaccines when the pandemic hit, and he quickly shifted to trials for most of the major COVID vaccine candidates (as well as some treatments). He even participated in one of the clinical trials himself, hoping that it would convince more people who look like him that the vaccines are safe. As a result, more people of color participated in those trials than in any previous one that Wallace had been involved in.

Even though vaccine trials had to be sped up during COVID—just as drug trials were—the standards for safety and effectiveness didn't change. Every vaccine that got emergency approval by the WHO was tested for safety in thousands of people around the world. In fact, because COVID vaccines have been given to so many people and their safety records tracked so closely, scientists now have extensive safety data on the various ones on the market—even in groups like pregnant women, who are usually not prioritized in clinical trials for vaccines because of the potential side effects for the babies they're carrying.

Another reason COVID vaccines got approved so quickly is that the people responsible for approving them worked insanely hard, reducing a process that can take years to a matter of months. Government employees in Washington, D.C., Geneva, London, and other cities worked around the clock to examine data from vaccine trials and go through those hundreds of thousands of pages of documents. Keep this in mind the next time you hear someone bad-mouthing government bureaucrats. If you're one of the people who got a COVID vaccine fast, and if you felt confident that it wouldn't seriously hurt you, you have many unsung heroes at the FDA working long hours away from their families to thank for it.

We will need to accelerate trials and approvals even more next time. The efforts that I mentioned in Chapter 5 to prepare for trials in advance—agreeing on protocols and setting up the infrastructure for running them, for instance—will help with vaccines just as much as with drugs. In addition, during COVID, researchers and regulators have learned a great deal about how safe mRNA and viral vector vaccines are, and they'll be able to use this understanding to evaluate candidates even faster in the future.

Let's continue the hypothetical outbreak example from Chapter 5. Suppose the world wasn't able to contain it in time, and it's going global, so we'll need to vaccinate billions of people. Several vaccines have made it through the approval and review process and are licensed for use in humans. Now we need to solve a whole other set of problems: How do we make enough of them and distribute the doses so they do the most good?

To give you a sense of how many more doses we'll need to make: The world typically produces 5 to 6 billion doses of vaccines every year—that's counting all childhood vaccinations, flu shots, polio vaccines, and more. If there's a huge outbreak, we'll need to make nearly 8 billion doses of a new vaccine (one for just about every person on earth), and maybe as many as 16 billion (if it's a two-dose vaccine). And we need to do it without backsliding on other lifesaving vaccines. Our goal should be to do it in six months.

And, if you're a vaccine manufacturer, you're going to face challenges at each step in the manufacturing process:

- In the first step—producing the active ingredients that make your vaccine work—you might need to grow cells or bacteria, infect them with the pathogen you want to stop, and harvest the substances they produce for your vaccine. To do this, you'll need a container called a bioreactor—either a reusable steel vat or a single-use plastic bag. But the supplies of both are limited. Early in the pandemic, bioreactors were bought up en masse by various companies hoping for an early breakthrough. If you couldn't find toilet paper at the store, you know how they felt.
- Next you'll mix your vaccine with other things to make it more effective or stable. If it's an mRNA vaccine, you'll need a lipid to protect the mRNA. If it's another type, you might need an adjuvant. Unfortunately, Chilean soapbark trees can be hard to come by, so if you want to use their bark as your adjuvant, your manufacturing may be held up while you wait for your allotment. In the future, we need

to be able to make more synthetic versions of adjuvants so we can scale up manufacturing quickly.

- Finally you'll need to pour your doses into vials. You'll need sterile, highly precise equipment, and the vials will have to meet exacting specifications, down to the type of glass and the stopper. (At one point during COVID, there were fears that the world would run out of the high-grade sand used to make this glass.) You'll need to label your vials according to rules set by the country in which you'll sell your vaccine, including which languages can be used, and those rules will vary from place to place.

For years there has been a debate in global health about when waiving companies' intellectual property rights is an effective way to get more vaccines or drugs produced. In some cases, waivers have helped make low-cost drugs available, as with the HIV treatments I mentioned in Chapter 5. This history got a lot of attention again in 2021, as advocates called on the World Trade Organization to waive the intellectual property protections for COVID vaccines.

The world absolutely needed to make more vaccines, and there are ways to accomplish that, which I'll come to later in this chapter. Unfortunately, the calls to waive intellectual property came too late to help close the supply gap. There are a limited number of facilities and people in the world who can produce vaccines that meet all the national and global requirements for quality and safety. And because most vaccines are made using very specific processes, you can't simply switch a facility from making, say, viral vectors to making mRNA vaccines. You'll need new equipment and training for your workers, and even then, your facility will still need to be approved to make the new product.

Suppose companies were required to release their recipes for approved vaccines, and now Company B wants to duplicate Company A's approved vaccine and meet all the necessary standards. But just getting A's recipe won't be enough—they'll also need information from A, like specifics about their manufacturing processes, data from clinical trials, and details about what they've told regulators. Because some of this information applies to A's other products—maybe they want to use the same process to make a cancer vaccine—they will hesitate to release it.

If B proceeds anyway but deviates from A's manufacturing process in even the slightest way, they'll have to go through new clinical trials, defeating the purpose of getting A's recipe in the first place. And in the end, the two companies will release two seemingly similar products that may have different levels of safety and effectiveness—creating confusion about vaccines at a time when what everyone needs is clarity. Company B benefits because it can't be sued by A, but otherwise it has not gained much.

What's more, making a vaccine is typically more complicated than making a drug. Remember that many drugs are made using chemical processes that are well defined and measurable, but many vaccines don't work that way. Manufacturing them often involves living organisms—anything from bacteria to chicken eggs.

Living things don't necessarily act exactly the same way every time, which means that even if you follow the same process twice, you might not get the same product both times, and it's much harder to tell whether the generic version matches the original in all the important ways. The process of making a vaccine typically involves thousands of steps! Even an experienced vaccine maker finds it difficult to duplicate another company's process and is most successful when it gets technical assistance from the original producers.

This is why we have generic drugs but not generic vaccines. Although the situation might change in the future, particularly as the mRNA vaccine technology matures, it's not a realistic possibility right now. Waiving intellectual property protections in 2021 would not have meaningfully increased the supply of COVID vaccines when we needed them.

The key decisions that determined how quickly there would be enough vaccine supply for the entire world were made in 2020. In the first half of that year, several organizations—including CEPI, Gavi, national governments, and the Gates Foundation—worked with many of the companies in the vaccine ecosystem on arrangements that would maximize the number of vaccines produced. The approach used wasn't simply to make the intellectual property available and ask manufacturers to do their own factory design and trials, but rather to cooperate and share all the information—including factory design and methods for verifying the quality of the vaccines—and work together with regulators. Until 2020, such deals were rare, but given the urgency of producing lots of vaccines quickly, they were the best way to get additional factories into production without compromising regulatory approval or quality.

These deals are known as second sourcing. In a second-source deal, one company with a viable candidate agrees to work with another company to make that vaccine in its own facility. They share not only the recipe but also the knowledge about how to use it, as well as personnel, data, and biological samples. Imagine buying a cookbook by David Chang and then he shows up at your house, along with the ingredients, to walk you through his ramen recipe.

These are complex arrangements that involve accounting for the costs and time associated with the transfer, negotiating the necessary licenses, and settling on terms that will be acceptable to both parties. And there are lots of incentives pushing against them for both companies: Imagine Ford inviting Honda to use its factories to make Accords.

But when they work, these arrangements are remarkable. It was second-source deals—not a government-mandated intellectual property waiver—that allowed Serum Institute of India, or SII, to produce one billion doses of COVID vaccine at a very low cost and in record time.

Before COVID, most vaccines destined for low- and middle-income countries were created not via second-source agreements, but by low-cost manufacturers who got philanthropic funding to do some of the development on their own. But during the pandemic, companies made more second-source deals than ever before. In less than two years, a single manufacturer, AstraZeneca, signed second-source deals involving twenty-five factories in fifteen countries. (Recall that AZ also agreed to forgo its profits on the COVID vaccine.) Novavax also signed one with SII—leading to a vaccine now being used in many countries—and Johnson & Johnson signed one with the Indian company Biological E. Limited and the South African firm Aspen Pharmacare. All told, second-source deals led to the production of billions of additional COVID vaccine doses. And in the future, such deals could be done even faster if companies that have them now can maintain their relationships with one another, so they don't have to start from square one during the next outbreak.

I'm also hopeful that this is yet another problem that mRNA vaccines will help solve. Many of the conventional ways to make vaccines are pretty arcane, which means there are many *i*'s to dot and *t*'s to cross in making a second-source deal. But because the basic approach to mRNA is pretty much the same—you just swap out your old mRNA for the new one and make sure the lipid is made the right way—it should be easier to transfer between companies. There are also some new modular technologies in the pipeline that, if they prove out, will make it cheaper and easier to build and run factories. This would make them more flexible, able to adapt as needed to make different types of vaccines.

Finally, there are a couple of steps that global bodies like the WHO and CEPI can take. The WHO should standardize the labels that go on vials, so companies don't have to make bunches of different labels for the same vaccine. CEPI and others should buy up raw materials for making vaccines and vials ahead of time, and distribute them later to the manufacturers with the most promising candidates. CEPI did this with glass vials during COVID, ensuring that there would be a reserve supply if any companies hadn't acquired enough on their own.

COVID vaccines significantly reduce the risk of severe disease and death, but how quickly you got one depended in large part on whether you live in a rich country, a middle-income one, or a poor one. In 2021, more than half of the world's population received at least one dose of a COVID vaccine. In low-income countries, only 8 percent did—and even worse, young, healthy people in rich countries who aren't likely to get sick or die from COVID were getting vaccines before older people and frontline workers in poorer countries, who were at much higher risk.

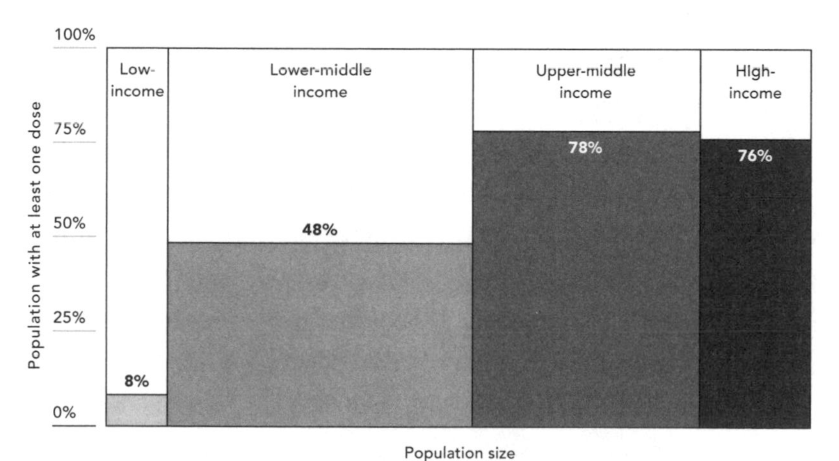

Vaccine inequity. As of December 2021, people living in wealthier countries had much higher COVID vaccination rates than people living in lower-income ones. The width of each box represents its share of the global population. (Our World in Data)

In theory, we could have reduced these inequities by allocating more fairly the doses we have on hand. It's true that rich countries have fallen short of their commitments to share more than a billion doses with poorer countries during COVID, but meeting those commitments wouldn't have been enough to fill the gap. And sharing doses in itself is not a permanent solution—there's little reason to think that rich countries will be more willing to do that in the future. How many politicians will tell young voters they can't be vaccinated because the doses are going to another country, at a time when schools are still closed and people are still dying?

That's why I think that, rather than focusing primarily on reallocation, the more realistic approach is to focus on making more doses—so many that the question of who should be able to draw from a limited supply is no longer an acute issue. In 2021, the White House published a smart plan with ambitious goals: Develop, test, manufacture, and distribute a safe, effective vaccine to everyone in the world within six months of recognizing a threat. If it's a two-dose vaccine, that means producing some 16 billion doses roughly six months after identifying the pathogen.

So let's go through what it will take to make enough doses for the world, starting with a look at how vaccines are priced and the methods for making them cheaper.

Because it costs so much to develop a new product, the companies that invent new vaccines try to recoup their costs as quickly as possible by selling doses at the higher prices that rich countries can afford. Even if their original production process results in a vaccine that's quite expensive, they have little incentive to redesign it, because then they would have to go through a new regulatory review.

For a number of vaccines, the solution has been to work with manufacturers in developing countries to come up with a new vaccine for the same disease, while also making sure that the cost of producing it is very low. This is far easier than inventing the vaccine in the first place, because you know it's possible to do it, and you understand the immune response that you need to provoke.

The pentavalent vaccine—which protects against five diseases—is a great example. The most widely used one was invented in the early 2000s, but there was only one manufacturer, and at more than $3.50 per dose, it was quite expensive for low- and middle-income countries. The Gates Foundation and its partners worked with two vaccine companies in India—Bio E. and SII, the same two that more recently started producing COVID vaccines—to develop a pentavalent vaccine that would be affordable everywhere. These efforts drove the price to below $1 per dose, and they drove up coverage levels so that more than 80 million infants get three doses of the vaccine every year. That's a sixteen-fold increase in coverage since 2005.

Similar deals have led to new vaccines for two major killers of children: rotavirus and pneumococcal disease (a severe respiratory illness). Both Serum Institute and Bharat Biotech, which is also based in India, created affordable rotavirus vaccines that are now available to every child in that country. They're used in several African countries too, and both companies are trying to make them even easier to administer in the world's poorest nations. And, as I was writing this book, India announced that it would expand access to the pneumo vaccine from less than half the country to the entire country—a decision that will save the lives of tens of thousands of children every year.

The Gates Foundation has been the largest funder of vaccine manufacturing in developing countries over the past two decades. What we've learned from that experience is that, for these countries, creating an entire vaccine-making ecosystem is a long walk down a hard road. But the obstacles can be overcome.

For one thing, there's the issue of regulatory approvals. The WHO has to approve every vaccine purchased by U.N. agencies such as COVAX. If the vaccine is first approved by the United States, the European Union, or one of a handful of other governments around the world, then the WHO's review is relatively quick. Otherwise, the WHO's review will be much more thorough and could take as long as a year (though the organization is trying to speed up the process for all approvals).

India and China, both of which have robust vaccine-manufacturing industries, are working to get a designation that would make the WHO's reviews even faster. Once they have it, the vaccines and other innovations made in those countries will be usable in the rest of the world even sooner than they are now. In Africa, regional groups are working with the WHO and other partners to improve the quality of regulation on the continent, and governments have begun to adopt international standards for vaccines so manufacturers don't have to meet different requirements in each country.

There's another challenge in addition to the approval process: Vaccine manufacturers need to make other products between outbreaks, or they'll go out of business. As new vaccines become available for diseases like malaria, tuberculosis, and HIV, they'll increase the overall size of the vaccine market, potentially making room for new producers. And in the meantime, countries can take on the fill-and-finish process—putting vaccines made elsewhere into vials and distributing them.

In the mid-2000s, during a trip to Vietnam, I visited a rural health clinic, hoping to get a firsthand look at some of the challenges the staff there were dealing with. As a big fan and funder of vaccines, I was especially interested in what it took to deliver a vaccine over what people in the field call "the last mile"—the journey from a storage facility to a remote clinic and finally to a patient.

The clinic had just received a shipment of the new rotavirus vaccine that I mentioned in Chapter 3, but there was a problem. To make a point, one of the health workers took a few of the vials and tried to put them in a portable cooler. (These coolers are what vaccinators carry vaccines in when they go into the field.)

The new vials didn't fit in the cooler.

This may seem like a minor detail, but it's a big problem. Most vaccines will become ineffective if they aren't kept cold—typically between 2 and 8 degrees Celsius, or roughly 35 to 46 degrees Fahrenheit—during the trip from the factory to their final destination. If a health clinic can't keep the vials cold, the doses will not be effective and will need to be thrown out. (Keeping them at the right temperature throughout the process is known as *maintaining the cold chain.*)

The manufacturer of the rotavirus vaccine soon fixed the problem by changing the size of the vials, but it was a vivid illustration of a fundamentally important point about vaccines: Delivering them to every part of the world where they're needed is an enormous logistical challenge, and seemingly small decisions like the size of a container can throw everything off.

The good news is that the cold chain and other hurdles in delivering vaccines have been solved in most parts of the world. Today, 85 percent of children get at least three doses of the pentavalent vaccine I mentioned earlier. But getting to the remaining 15 percent is a big challenge.

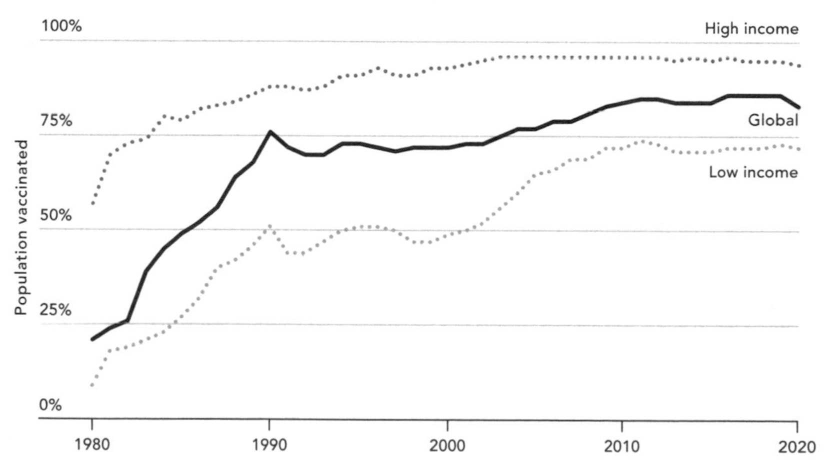

Global vaccination rates are higher than ever. The percentage of children who receive three doses of the vaccine that prevents diphtheria, tetanus, and pertussis (DTP3) has risen dramatically since 1980. (WHO)

For the sake of making sure every child gets the basic vaccines that most of the world enjoys, and preparing to stop outbreaks before they go global, we need to be able to deliver vaccines everywhere, even in the most remote places. Let's look at what it takes to get a vaccine from the factory all the way to the patient.

Depending on where a container of vaccines is headed, there

might be as many as seven way stations along the route. The container arrives in a country by boat or plane and is taken to a national storage facility. From there, it will be shipped to a regional facility, then a district one, then a subdistrict one, and then a community facility. A health worker will load up a pack of vaccines and go to remote areas, vaccinating people in and around their homes.

At every step of the way, the container has to maintain just the right temperature, not just at each storage facility, but on the road from one facility to another as well. The power could go out at any one of the facilities, shutting down refrigerators and putting the vaccines at risk of losing their effectiveness. Pfizer's mRNA vaccine has to be stored at minus 70 degrees Celsius, or minus 94 Fahrenheit—a major difficulty for developing countries, where keeping vaccines cold was already a challenge.

Ultimately, vaccines reach the people who need them only because of the health workers who are dedicated to taking them that last mile. Their work requires precision and stamina—they often have to walk many miles every day on their rounds—and can be risky. Depending

A health worker in Nepal travels several miles each day, often through rigorous terrain, to deliver vaccines to people living in remote locations.

on which vaccines they're administering, they might have to prepare each dose by diluting a powder in a liquid, carefully using just the right proportions of each. They might get stuck by a needle during the process. They have to watch out for counterfeit vaccines. They must keep good records of who has been vaccinated.

There's phenomenal work going on to solve these problems. Auto-disable syringes have a built-in safety mechanism so you can't accidentally poke yourself with one or use it more than once. These syringes have literally been lifesavers in vaccinating children against pertussis and other diseases, but during the pandemic, there was so much demand to use them for COVID vaccines that it put routine childhood immunization programs at risk. UNICEF and other organizations stepped in to get more auto-disable syringes made and distributed.

Vaccinators in India are using a new portable cooler that prevents the vaccines from freezing if the ice in the cooler is too cold. Researchers are also working on new vaccine formulations that wouldn't need to be kept as cold at every step of the journey. They're reducing shipping costs and saving refrigerator space by making the packages smaller, and they're simplifying the process for health workers by getting rid of powders that have to be mixed with liquid on site.

Bar codes printed on the vials will let vaccinators use their cell phones to confirm that the vaccines are legitimate, just as you might scan a QR code to pull up the menu at a restaurant. When each vial is scanned, health officials can keep track of exactly how many have been used, which will tell them when the clinic is running low and needs to be resupplied. Advanced methods for delivering vaccines, such as replacing the needle and syringe with a small patch containing microneedles—picture something that looks superficially like the nicotine patches that people use to stop smoking—could make the process safer for everyone, and they may make vaccines easier to deliver too.

—

You've probably read (in this book if not somewhere else) that the main goal of vaccines is to prevent severe sickness and death, not to prevent infection. But that's not ideal, of course: A perfect vaccine would in fact keep you from getting infected, which would make a big difference in cutting down transmission—no one who was vaccinated could pass the pathogen to other people. The measles vaccine is a good example: After two doses, it gives you 97 percent protection against infection.

Getting other vaccines to that level is a long-term goal, and one especially promising route is to administer them in different ways, to different parts of your body. Think about how you contract COVID—the virus enters your body through your nostrils and airways, where it attaches to your mucus. But getting a vaccine jabbed into your shoulder doesn't generate much immunity in your mucus cells. To do that, it will probably be better to have vaccines that you inhale as a nasal spray or swallow in liquid form.

Humans have antibodies that are specialized for the wet surfaces of our nose, throat, lungs, and digestive tract. These antibodies have more sites where they can grab on to a virus than the antibodies in your blood have, which makes them more efficient virus hunters. (One unpublished paper I've seen suggests that, in mice at least, these cells may provide ten times as much protection.)

In the future, you may be able to inhale or swallow a vaccine that gives you immunity inside your body to prevent severe infection or death, and immunity on your mucosal surfaces too—which would protect you and reduce the odds that you'll transmit a virus through breathing, coughing, or sneezing. When Larry Brilliant and other scientists were asked to come up with an imaginary vaccine for the hypothetical virus depicted in the movie *Contagion,* they chose a nasal spray—because, they later wrote, "it would be easy to manufacture worldwide, distribute and deliver."

In addition to these new methods of delivering vaccines, we should also pursue another possibility: infection-blocking drugs that can be combined with vaccines. The drug would provide short-term protection from infection, and the vaccine would act as a backstop, giving you long-term protection from severe disease. You would use the drug when the disease was spreading especially fast, but if it didn't work or you didn't take it often enough, you would still have the vaccine working to keep you out of the hospital.

The technology behind these drugs is still in its early days, but if we could get them to the point where they could be developed quickly—the way mRNA vaccines now can be—and delivered by nasal spray or pill, they'd be a phenomenal tool for keeping an outbreak at a low level.

And if they were cheap and long-lasting enough—a few pennies for a dose that lasts thirty days or more—it might make sense to use them to block seasonal respiratory infections. Every schoolchild could get a dose at the beginning of each month. You could even set up sniffing stations where people would stop by every few weeks for another dose.

There's some exciting work being done on this category, which I'll call blockers. The company Vaxart, for example, has produced promising data on an oral blocker for flu and is working on one for COVID. On the whole, though, this approach isn't getting nearly enough attention, given what a breakthrough it would be for new and existing diseases alike. Governments and companies need to invest a lot more in it, with a particular focus on making it affordable and practical in low-income countries as well as rich ones.

None of these tools will matter, though, if people refuse to use them. Whenever I talk to someone about blockers or vaccines, whether it's a scientist, politician, or journalist, there's one subject hanging over everyone's heads: vaccine hesitancy. I imagine that someday we'll have to confront blocker hesitancy too.

Researchers who are studying vaccine hesitancy have gained a few insights. One is that there isn't one single reason for it. Fear and suspicion are certainly factors. So are things like how much people trust the government and their ability to get information that's timely and accurate. Many Black Americans, for example, are generally skeptical of the government's good intentions when it comes to health, and understandably so. For forty years, the U.S. Public Health Service ran the infamous Tuskegee Study—a horrific experiment in which it looked at the effect of syphilis on hundreds of Black men, without giving them their true diagnosis, and even withholding treatment once it became available eleven years into the study.

There are also socioeconomic factors that don't have anything to do with fear, mistrust, or misinformation—like whether you're able to get to an immunization site. Many people don't have the transportation to get to a clinic that's miles away. Maybe they can't afford to stop working or get someone to watch their kids. Safety is also a consideration for women who need to travel long distances by themselves to get the vaccine.

But over the years I've learned that you can't persuade uncertain people simply by throwing more facts at them. You need to meet them where they are—literally and figuratively.

That means vaccines need to be affordable or free, and available nearby at a time when people can get to them. It can help when people see politicians and celebrities getting vaccinated. And maybe most of all, they need to hear the truth from trusted sources, like religious leaders and local health workers they already know.

In Zambia, anyone in search of good information can tune the radio to FM 99.1. Once a week, Catholic nun and social worker Sister Astridah Banda hosts the "COVID-19 Awareness Program," a talk show where she and her guests discuss health topics—with a focus on preventing COVID—and answer questions from callers. Sister Astridah is not a doctor, but she is passionate about public health. When COVID arrived in Zambia, she noticed that most of the public health bulletins were written in English. And although

English is an official language in Zambia, many people speak only one of the country's local languages and were missing out. She approached Yatsani Community Radio and asked to start broadcasts on which she could translate the bulletins into local languages and share other information about the virus. Her show now reaches more than 1.5 million people.

Sister Astridah Banda, a nun and social worker, passes along information about COVID on Yatsani Community Radio in Lusaka, Zambia.

In any outbreak, the world needs many Sister Astridahs, and other efforts too. Driving up vaccination rates requires both supply and demand—you need to have enough vaccines, and people need to want them. As I've argued in this chapter, innovative policies and technologies will help us make and deliver enough vaccines for everyone. Making sure the demand is there too will be just as important.

This chapter boils down to two key points. First: As awful as COVID is, the world is lucky to have made vaccines as quickly as it did. And second: We've only touched the surface of how good vaccines can be. Because we can't assume we'll be so lucky next time—and because there are phenomenal opportunities for saving lives beyond the threat of pandemics—the world should be pursuing an ambitious agenda to make vaccines even better.

I see six areas that should be priorities for funding and research:

- **Universal vaccines.** Thanks to the advent of the mRNA vaccine, it should be possible to create injections that target several variants of the same pathogen, or even multiple pathogens. We could have vaccines that protect you from coronaviruses, influenza, and the respiratory virus known as RSV—and with some luck, we could even eradicate all three virus families.
- **One and done.** One of the big hurdles in COVID vaccination has been the need to deliver multiple doses. It's an inconvenience for people who can get to a clinic or pharmacy easily, who don't have to worry about child care, and who can get time off work, but for others, it's a huge barrier. New formulations of vaccines would give you the same protection with one shot as you now get with two; given the work that's already under way, I think this is an achievable medium-term goal. And an ideal vaccine would protect you for your whole life rather than requiring yearly boosting; research into the immune system should allow us to figure out how to provide such long-lasting protection.
- **Total protection.** The best COVID vaccines available (as I write this, at any rate) cut your risk of infection, but they don't eliminate it. If we can make vaccines that give you total protection, then we would shut down the spread of

the disease substantially—breakthrough cases would be a thing of the past. We need to generate more protection in your mucosal tissues, including the mouth and nose.

- **No more ice chests.** Vaccines would be far easier to deliver, particularly in the developing world, if they didn't have to be kept cold all the time. Researchers have been working on this problem since at least 2003, and we still don't have a complete solution. If we did, it would revolutionize the field of vaccine delivery in poor countries.
- **So easy, anyone can give it.** Vaccines and infection-blocking drugs that you could take as a pill or inhale with a nasal spray would be much easier to administer than ones that have to be injected. And the micro-needle patches I wrote about earlier would make needles and syringes obsolete. You could pick one up at the grocery and apply it yourself, without help from a health care worker who has to put a needle in your arm, and it might not even need to be kept cold. Researchers are already testing prototypes that deliver measles vaccines, and although this work is moving ahead fast, we need more time and effort to get them ready for market, produce huge volumes of them, and use the patch technology as a platform for going after other diseases.
- **Expanded manufacturing.** For all these advances to have an impact, developing and approving them won't be enough. We'll also need to manufacture them in massive volume—enough for the entire world—and within six months. To do that, we'll need production capacity throughout the world, including in regions with the worst disease burden. And we'll need to be creative about how all this new infrastructure can stay in business even when there isn't a pandemic threatening to emerge.

CHAPTER 7

PRACTICE, PRACTICE, PRACTICE

In July 2015, *The New Yorker* published an article that got attention up and down the West Coast of the United States. I live just outside of Seattle, and I remember emailing the article to friends just as it was arriving in my inbox from other friends. It became a regular source of dinner-table conversation that summer.

The headline on the article was "The Really Big One: An earthquake will destroy a sizable portion of the coastal Northwest. The question is when." The author, a journalist named Kathryn Schulz who won the Pulitzer Prize for this piece, explained that a huge stretch of coastline, from Canada into Washington state, Oregon, and northern California, sits near what's known as the Cascadia Subduction Zone. Cascadia is a fault hundreds of miles long beneath the Pacific, where two tectonic plates meet and one is sliding underneath the other.

Subduction zones are inherently unstable and tend to cause earthquakes. Seismologists figure that massive earthquakes occur along the Cascadia zone an average of every 243 years, and that the last one occurred around 1700. The 243-year average is debated, and Cascadia may go much longer than that between quakes, but when we read the article, none of us locals could dismiss the fact that the last Cascadia quake happened more than 315 years ago.

The article cited horrific projections: A Cascadia quake, and the tsunami that would result from it, could kill nearly 13,000 people, injure 27,000 more, and displace a million people from their homes. And the toll could be much worse if the quake occurred during the summer, when tourists crowd the beaches of the West Coast.

To test how ready the Pacific Northwest is for the really big one, the federal government oversees a series of periodic full-scale exercises known as Cascadia Rising. The 2016 exercise involved thousands of people from dozens of government agencies, the military, nonprofits, and businesses. A lengthy after-action report detailed the results and issued a series of lessons learned during this real-life exercise. Among other things, the report noted, "Catastrophic response requirements are fundamentally different than any response we have seen before. . . . A *massive* response will be required." Another Cascadia Rising exercise is scheduled for summer 2022.

I wish I could report that Cascadia Rising has led to major changes and the Pacific Northwest is now as ready as it can be for a catastrophic quake. Unfortunately, that's not the case. For one thing, retrofitting all or even most of the region's buildings to make them seismically sound would be prohibitively expensive.

But the exercises are still worthwhile. At least the government is trying to get people to focus on the problem.

We tend to use words like *drill* and *exercise* interchangeably, but in the world of disaster preparation, they don't mean the same thing at all.

A drill is a test of just one part of a system—say, whether the fire alarm in your building works, and whether everyone knows how to get out quickly.

Then, moving up the scale of complexity, there's the tabletop exercise, a discussion designed to identify and solve problems. More complicated still is the functional exercise, a simulated disaster that tests how well the whole system operates, but without moving people or equipment.

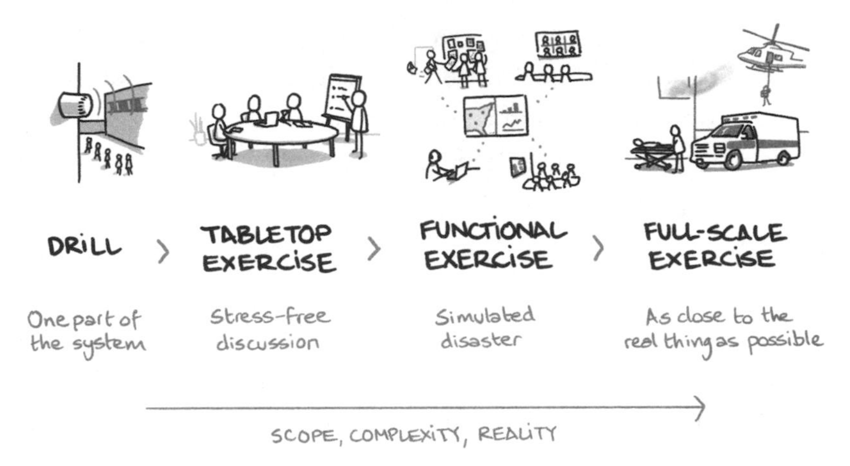

Finally, there are full-scale exercises, such as Cascadia Rising. These are designed to get as close to the real thing as possible—complete with actors pretending to be sick or injured and vehicles moving people and equipment around.

For as long as I've been learning about pandemic preparedness and prevention, I've been amazed that there isn't an ongoing series of full-scale exercises designed to test the world's ability to detect and respond to an outbreak. As the WHO's flu preparedness program put it in a 2018 guide to running outbreak exercises, "Considerable efforts and resources have been invested by countries around the world in developing national pandemic influenza preparedness plans and the capacities needed to respond to an influenza pandemic. However, to be effective, plans need to be tested, validated and updated periodically through simulation exercises."

There have been many tabletop and functional exercises for disease outbreaks, but perhaps only a handful of country-scale ones designed to simulate an outbreak of flu or coronavirus.* Credit for

* Exercises related to animal-borne diseases aren't unheard of. For example, four years after a disastrous outbreak of foot-and-mouth disease in 2001, the United Kingdom and five Nordic countries ran simulations to test their readiness.

running the first one seems to belong to Indonesia, which held a full-scale outbreak exercise in Bali in 2008. There have been no exercises involving entire regions throughout the world.

Although details are sometimes murky because governments make some of the results classified—especially for full-scale exercises—it appears that the history of these simulations is spotty. On the positive side is Vietnam, which has held frequent simulations at various levels of complexity, taken action to fix problems that were revealed, and set itself up to respond especially well to COVID.

But often, in other countries, these exercises end with a series of what-ifs and missed opportunities.

For example, the United Kingdom ran an exercise called Winter Willow in 2007 and another, Cygnus, in 2016, both focused on flu outbreaks. Cygnus in particular highlighted problems with the government's readiness and produced a series of classified recommendations that went unheeded—and which caused a scandal when *The Guardian* revealed them during the first year of the COVID pandemic.

The United States had a similar experience in 2019, when the government ran Crimson Contagion, a series of exercises designed to answer one question: Was the country ready to respond to an outbreak of a novel flu virus?

Overseen by the Department of Health and Human Services, Crimson Contagion occurred in two phases. The first involved a series of seminars and tabletop exercises held between January and May, in which people from all levels of government, plus the private sector and nongovernmental organizations, got together to discuss existing plans for responding to an outbreak.

In the second phase, they put these plans to the test in a functional exercise. Over four days in August 2019, participants worked through a scenario in which tourists visiting China have become ill with a respiratory illness caused by a virus. They fly out of the airport in Lhasa and travel to other cities in China before flying home to their respective countries.

The virus turns out to be just as contagious as the 1918 flu strain, and only slightly less deadly. It rapidly spreads from one human to another, making its first U.S. appearance in Chicago and quickly moving to other major cities.

At the beginning of the exercise, it has been forty-seven days since the first U.S. case. There are moderate or high caseloads across the Southwest, Midwest, and Northeast. Models are predicting that the virus will make 110 million people in the U.S. sick, put more than 7 million in the hospital, and kill 586,000 Americans.

Over the next four days, the participants would debate decisions that would have been unfamiliar to anyone not already versed in outbreak response work: quarantines, personal protective gear, social distancing measures, school closures, public communications, the purchase and distribution of vaccines. Today, of course, these terms are part of our everyday vocabulary.

The scope of Crimson Contagion's functional exercise was enormous. It involved 19 federal departments and agencies, 12 states, 15 tribal nations and pueblos, 74 local health departments, 87 hospitals, and more than 100 groups from the private sector. When it was all over, the participants got together to discuss how it had gone. While they found a few things that worked well, they found many more that did not. I'll mention just a few, which will sound eerily familiar.

No one in the exercise understood what the federal government was responsible for, versus what others would do. The Department of Health and Human Services didn't have clear authority to lead the federal response. There wasn't enough money to buy vaccines (in this scenario, there was already an available vaccine for the strain in question, but it hadn't been administered). State leaders didn't know where to turn for accurate information. There were huge discrepancies in how states planned to deploy scarce resources, such as ventilators, and some had no plan at all.

Some of the troubles were almost comically mundane, like something out of the TV show *Veep.* Federal agencies confused participants by unpredictably changing the names of conference calls. Sometimes the name of the meeting used some unrecognizable acronym, so people wouldn't show up. State governments, already understaffed, struggled to keep up with all the calls while also handling the response itself.

It is telling that in the official government report on the results of Crimson Contagion—dated January 2020, just as COVID cases were starting to mount—the word *diagnostics* appears only three times over fifty-nine pages. The report simply notes that diagnostics will be one of many supplies that will be hard to come by in a pandemic. Just a few weeks later, of course, the United States' inability to ramp up testing in a serious way would become tragically apparent. It bears repeating: America's failure to test people at anything near the level that other nations achieved is among the biggest mistakes any country has made during the pandemic.

Crimson Contagion was not the first simulation designed to test America's readiness to handle an outbreak. That honor likely goes to a tabletop exercise with the ominous name Dark Winter, which was held over the course of two days in June 2001 at Andrews Air Force Base in Washington, D.C.

Surprisingly, Dark Winter wasn't organized by the federal government, but by independent organizations whose leaders were growing increasingly concerned about the potential of a bioterror attack on the United States and wanted to draw attention to the problem.

Dark Winter supposed that a terrorist group releases smallpox in Philadelphia, Oklahoma City, and Atlanta, infecting a total of 3,000 people. Less than two months later, the disease had spread to 3 million people and killed one million, with no end in sight. An observer I know commented that the outcome was Smallpox 1, Humanity 0.

Other exercises followed: Atlantic Storm in 2005 (another smallpox attack), Clade X in 2018 (an outbreak of a novel influenza virus), Event 201 in 2019 (an outbreak of a novel coronavirus), a simulation at the Munich Security Conference in 2020 (a bio-attack involving an engineered influenza virus).*

Even though each of these U.S. exercises imagined different scenarios, were run in different ways, and used different methods, they had three things in common. One is that their conclusions are fundamentally the same—there are huge gaps in America's and much of the world's ability to contain outbreaks and prevent a pandemic—and they propose various ways to fill those gaps.

The second thing these exercises have in common is that none of them led to any significant changes that would make America better prepared for an outbreak. Although some adjustments were made at the federal and state levels, we only need to look at what's happened since December 2019 to see that whatever changed was not enough.

The third is that, with the exception of Crimson Contagion, each of the U.S. simulations took place exclusively in conference rooms, and none of them involved moving real people or equipment from one place to another.

Full-scale exercises aren't run as often as tabletop and functional ones for the obvious reason that they're expensive, time-consuming, and intrusive. In addition, some public health leaders have argued that the best way to prepare for a pandemic is to simulate smaller outbreaks, which means not preparing for things that happen only in an epidemic or pandemic—problems like supply chains being disrupted, economies shutting down, and heads of state interfering

* The Gates Foundation was one of the funders of the Event 201 exercise. Some conspiracy theorists suggested that it predicted COVID. As the organizers made clear, it wasn't a prediction, and they said so at the time. You can find a statement about this at centerforhealthsecurity.org.

for political reasons. It's also likely that, until 2020, the threat of a worldwide contagion seemed remote to most people, and therefore not worth the trouble and cost of a full-scale, real-life exercise.

Two years into COVID, the argument is much easier to make. The world needs to be running far more full-scale exercises that test its readiness for the next major outbreak.

In most countries, these exercises can be run by national public health institutions, emergency operations centers, and military leaders, with the GERM team that I described in Chapter 2 acting as an advisor and reviewer. For low-income countries, the world will have to bring in resources to help out.

Here's how a full-scale outbreak exercise might work. The organizers would pick a city and act as if it's experiencing a bad outbreak that could spread nationally or globally. How quickly can a diagnostic test for the pathogen be developed, manufactured at a high volume, and delivered wherever it is needed? How well, and how quickly, can the government get accurate information out to the public? How do local health officials handle quarantines? And—as we now know can happen—what if supply chains are cut, local health agencies make poor decisions, and political leaders interfere?

They would establish a system for reporting cases and running genetic sequencing on the pathogen. They would recruit volunteers to try out nonpharmaceutical measures, modify them based on how the disease spreads, and understand the economic impact they would have during a real emergency.

And if the pathogen initially spreads through human contact with animals, the exercise would evaluate a government's ability to dispose of the animals.* Suppose it's an avian flu spread by chickens:

* In November 2020, the Danish government ordered the culling of 15 million mink out of concern about a COVID mutation that might move from them to humans.

Because so many people rely on chickens for their livelihood, they'll be reluctant to slaughter the birds on the off chance they might spread a flu. Does the government have the money to compensate them for their losses, and a system for doing so?*

To make the exercise even more realistic, software would generate surprise events from time to time, throwing a wrench into the plan to see how everyone responds. Software would also be used to track the overall simulation and to record actions for later review.

In addition to advising countries on their simulation plans, the GERM team would measure readiness in other ways—for example, by looking at how well a given country's health system is detecting and responding to nonpandemic diseases. If it's a place where malaria is a problem, how early does the system detect big outbreaks? Or with tuberculosis and sexually transmitted diseases, how well does it trace the recent contacts of people who test positive? On their own, these proxies wouldn't tell researchers everything they need to know, but they would offer clues about weaknesses in the system that need more attention. The countries that do a good job of watching for, reporting, and managing endemic diseases are in a good position to respond to a pandemic threat.

The GERM team's most important role will be to distill the findings from exercises and other measures of preparedness, record the recommendations that come out of them—ways to strengthen supply chains, better methods for coordinating across governments, agreements to improve the distribution of medicines and other supplies—and then try to keep the pressure on world leaders to translate these findings into action. We've already seen how little

* If you want more detail from experts on what an exercise might involve, see the WHO document "A Practical Guide for Developing and Conducting Simulation Exercises to Test and Validate Pandemic Influenza Preparedness Plans," available at who.int.

things changed after Dark Winter, Crimson Contagion, and the other outbreak simulations. Unfortunately, there's no innovation that can make sure that after-action reports don't simply get stored on some website and then forgotten. Political leaders and policy-makers will need to change this.

To get a sense of the different scales at which full-scale exercises can be run, let's look at two examples from disaster preparedness, starting with a relatively small one.

In the summer of 2013, Orlando International Airport in Florida simulated a horrific aviation-related disaster, an exercise designed to meet the federal government's requirement that all U.S. airports run a full-scale simulation once every three years. In the scenario, according to an article in *Airport Improvement* magazine, a hypothetical jetliner carrying ninety-eight passengers and crew members experiences hydraulic problems and crashes into a hotel a mile from the airport.

The exercise involved 600 volunteers pretending to be victims, 400 first responders, and staff from sixteen hospitals, and it took place in a training facility with three aircraft and a four-story building designed to let firefighters practice on real fire. Officials had to establish who was in command. First responders had to triage patients, treat the ones they could, and transport others to the hospital. Security had to manage a crowd of observers. Friends and family of the victims had to be notified. News reporters needed updates. The exercise identified some necessary improvements and cost about $100,000.

At the other end of the complexity spectrum, there's the full-scale exercise run by U.S. military forces in August 2021. Over the course of two weeks, personnel from the Navy and Marine Corps participated in the largest naval training event in a generation. The name of this exercise—Large-Scale Exercise 2021—understated its scope. Simulating concurrent wars with two world powers, LSE

2021 spanned seventeen time zones and involved more than 25,000 personnel, using virtual reality to allow participants to join remotely and to link units from around the world so they could share information in real time.

The analogy between war games and germ games is not perfect. Stopping an outbreak, after all, is different from fighting a war. Countries should be working together, not against one another. And unlike military exercises, outbreak simulations can involve the public and be highly visible so they're no more out of the ordinary than a fire drill.

Still, the ambition of the LSE is impressive. The exercise created the opportunity for organizations spread around the world to share data and make fast, informed decisions together. It is hard to read about that and not think: *We need something like this for pandemic prevention.*

A good model of a simulation is a full-scale exercise developed by Vietnam in August 2018, designed to see how well the system identified a potentially worrisome pathogen. I'm impressed by how meticulous it was.

Four actors were hired to play patients, family members, and their contacts, and they were given scripts with key information for the medical staff (who knew they were participating in an exercise). On Day 1, the actor portraying a fifty-four-year-old businessman arrived at the emergency room of a hospital in the northeastern province of Quang Ninh, complaining of a dry cough, fatigue, muscle pain, and shortness of breath. The doctor questioned him thoroughly enough to discover that he had recently traveled to the Middle East, where he could have picked up the MERS virus—a fact that, in combination with his symptoms, was enough to get him admitted to the hospital and isolated.

News of the worrisome case made its way up the chain of

command within minutes, and soon the members of a rapid response team had arrived at the hospital and at the man's residence. The actors were tested using throat swabs, which were then replaced with samples that had been spiked with the virus that causes MERS. Although the samples weren't actually driven to a lab, the organizers waited the length of time it would have taken to transport them before lab staff ran real tests and correctly identified the positive MERS cases.

The exercise didn't go off flawlessly—the organizers noticed a number of gaps in the process—but it would be surprising if it had been flawless. The point is that the gaps were identified and, most important, fixed.

This full-scale exercise was small, compared with the national and regional ones that the world needs, but it had many of the necessary components. If exercises like this one were run by more countries and in more regions, they would keep us from making a classic mistake: preparing for the last war.

It will be tempting to assume that the next major pathogen will be as transmissible and as lethal as COVID, and as susceptible to innovations like mRNA vaccines. But what if it isn't? There is no biological reason why the next pathogen couldn't be far more lethal. It could quietly infect millions of people before a single person starts feeling sick. Our bodies might not be able to knock it out with neutralizing antibodies. With germ games, we'll be able to test against the wide range of pathogens and scenarios that the next outbreak might present.

Since the risk of a pandemic is higher than the risk of an all-out war, we should be running an LSE-sized global exercise organized by the GERM team at least once a decade. Each region should run another major exercise in the same decade, with advice from GERM, and countries should undertake smaller simulations with their neighbors.

There is one reason to hope that the reports generated by future exercises won't be ignored: experience. In the early days of COVID, many experts thought the countries that had gone through the SARS outbreak in 2003 were better prepared for this pandemic. Having experienced how bad it was, the theory went, they were ready—politically, socially, and psychologically—to do what it took to protect themselves. The theory proved true. The places hit hardest in 2003 included mainland China, Hong Kong, Taiwan, Canada, Singapore, Vietnam, and Thailand. When COVID emerged, most of these places responded quickly and decisively, limiting the numbers of COVID cases for more than a year.

Maybe Crimson Contagion, Dark Winter, and the rest didn't have more impact because their scenarios seemed so remote at the time—at least to most people and most politicians. Now, though, the idea of a virus spreading around the world, killing millions of people and doing trillions of dollars in damage, is very real for all of us. We should take the outbreak of a disease at least as seriously as we take earthquakes and tsunamis. To keep a pandemic like COVID from happening again, we need to practice stopping pathogens early, learn which parts of the system need to improve, and be willing to change even when it's difficult to do so.

So far in this book, I've stuck to writing about naturally occurring pathogens. But there is another, even more unsettling scenario that disease exercises must account for—a pathogen that's intentionally deployed with the goal of killing or maiming huge numbers of people. In other words, bioterrorism.

The history of turning viruses and bacteria into weapons stretches back centuries. In 1155, Frederick I, the Holy Roman Emperor, laid siege to the town of Tortona (in modern-day Italy) and is said to have poisoned the local water wells with dead human bodies. More recently, in the eighteenth century, British soldiers distributed blankets used by smallpox patients to Native Americans. In the 1990s, members of the Aum Shinrikyo cult released sarin gas in the Tokyo subway, killing thirteen people, and reportedly released botulinum toxin and anthrax four times without causing any casualties. And in 2001, a series of attacks using anthrax sent through the U.S. mail left five people dead.

Today the natural pathogen that would make the most fearsome weapon is surely smallpox. It's the only human disease that has ever been eradicated from the wild, though samples are still kept in government labs in the United States and Russia (and possibly in other countries too).

What makes smallpox especially scary is that it spreads fast through the air and has an extremely high mortality rate, killing around a third of everyone who's infected. And because most vaccination programs stopped after it was eradicated in 1980, almost no one is immune to it any longer. The United States does have a stockpile of smallpox vaccines large enough to protect everyone in the country, but as we've seen with COVID vaccines, distributing the doses would not be a simple matter—especially when people are panicking about an attack—and it's unclear how the rest of the world could be protected.

Part of the risk stems from the fall of the Soviet Union. As my friend Nathan Myhrvold notes in his paper "Strategic Terrorism," an international treaty banned bioweapons in 1975, but the Soviet Union continued its program into the 1990s—"thereby producing thousands of tons of weaponized anthrax, smallpox, and far more exotic biological weapons based on genetically engineered viruses."

The chances that terrorists will get their hands on these existing weapons are compounded by the fact that the science behind engineering pathogens is no longer the sole province of highly trained scientists working in secretive government programs. Thanks to the advances in molecular biology of the past few decades, students at hundreds of colleges and universities around the world can learn everything they need to know to engineer a biological weapon. And some scientific journals have published information that a terrorist could use to concoct a new pathogen, a practice that has led to vigorous debate about how to share research knowledge without adding to the risk.

We haven't yet seen a mass attack using an engineered bioweapon, but it is certainly not out of the question. In fact, during the Cold War, Soviet and American labs produced bioengineered anthrax that was resistant to antibiotics and evaded every vaccine. A nation-state or even a small terrorist group that developed smallpox resistant to treatment and vaccines would be capable of killing over a billion people.

A new pathogen could be designed that is highly communicable and lethal but doesn't cause symptoms right away. Such a pathogen would spread quietly around the world, perhaps for years, before arousing suspicion. HIV, which evolved naturally, works this way; although people can infect others very quickly after acquiring it, their health may not fail for nearly a decade, allowing the virus to go undetected while they spend years passing it to others. A pathogen that operated this way but didn't require intimate contact to spread, as HIV does, would be far worse than the AIDS pandemic.

"To put it in perspective," Nathan writes, a single attack that caused 100,000 casualties "would kill more people than were killed cumulatively in all terrorist actions by all parties throughout history. It might take anywhere from 1,000 to 10,000 typical suicide bombings to equal it." It is this scale of catastrophe—the kinds of events that can kill hundreds of thousands, millions, or even billions of people—that deserves far more attention than it gets.

Now, I am an optimistic person who's naturally inclined to focus on solutions. But even I have to admit that it's hard to write a list of responses that feels adequate to the threat of bioterrorism. Unlike a natural pathogen, an intentionally made disease can be designed to get around our tools of prevention.

All the things we need to do to prepare for a deliberate attack are a super-set of those we need to do to prepare for a natural one. Outbreak exercises can focus on attack scenarios and test our readiness. Better treatments and vaccines are important, no matter what the source of the pathogen is. Better diagnostics that return a result in thirty seconds would make it more practical to screen people at airports or public events, where we'd be most likely to see the spread of an engineered pathogen, and of course they would be extremely useful for everyday testing as well. Mass-scale genomic sequencing of pathogens will help in an ordinary flu outbreak and during an attack. Even if an attack never comes, we'll be glad to have all these tools available.

We also need some approaches that are specifically designed to deal with deliberate attacks. I am hopeful that we'll have devices in airports and other big gathering places that detect pathogens in the air and sewage, but the technology is still years away. The U.S. government made an attempt at a much larger-scale version of this approach in 2003 with a program called BioWatch, which placed devices designed to detect airborne anthrax, smallpox, and other pathogens in cities throughout the country.

Although BioWatch still operates in twenty-two states, it is widely regarded as a flop. Among other faults, it relies on the wind blowing in exactly the right direction and takes up to thirty-six hours to confirm a pathogen. Sometimes the detectors fail to work for the most basic reason: They get unplugged.

Regardless of whether air-sniffing machines have a future, the possibility of a bioterror attack is another reason why the world should be putting far more money and effort into research on detecting, treating, and preventing diseases that can go global. Given the national security implications of an attack and the chance that the number of casualties could reach into the millions, more of this research should come from defense budgets. The Pentagon's budget is roughly $700 billion a year, while the National Institutes of Health budget is about $43 billion a year. As far as resources are concerned, the Department of Defense operates on a whole other level.

Although I'm optimistic that science will deliver better tools for stopping outbreaks from any source, governments should also consider a defense that is as low-tech as they come: a reward. There's precedent for it—governments frequently offer to pay people for information leading to the arrest of criminals and terrorists. Considering the scale of damage that can be done these days, governments should be willing to pay quite a lot to informants who help thwart a bio-attack.

Regardless of what the final bioterror plan looks like, it will need to survive shifting political winds. In the early 1980s, while he was running the CDC, Bill Foege worked with the FBI on a program for detecting and responding to bioterrorism. The program included simulations of attacks using different diseases, to see how such attacks would work, as well as a defensive plan for each disease. Foege's successor, convinced that such an attack would never happen, closed down the program. If the U.S. and the rest of the world make a big investment in germ games and get the public's attention, it will be much harder for a single political appointee to get in the way of protecting people.

CHAPTER 8

CLOSE THE HEALTH GAP BETWEEN RICH AND POOR COUNTRIES

On the whole, the world's response to COVID has been exceptional. In December 2019, no one had heard of the disease. Within eighteen months, multiple vaccines had been developed, proven safe and effective, and delivered to more than 3 billion people, or nearly 40 percent of the earth's population. Humans have never responded faster or more effectively to a global disease. We accomplished in a year and a half something that normally takes half a decade or more.

Yet within these phenomenal numbers, there were, and are, startling disparities.

To begin with, the pandemic has not affected everyone equally. As you may remember from Chapter 4, Black and Latino third graders in the United States have fallen twice as far behind in their classwork as white and Asian American students have. Across every age group in the U.S., Black and Latino people and Native Americans are twice as likely to die from COVID as white people are.

The pandemic's overall impact was toughest in the world's low- and middle-income countries. In 2020, it pushed nearly 100 million people around the world into extreme poverty, an increase of about 15 percent and the first time in decades that the number went up. And only a third of low- and middle-income economies are expected to return to pre-pandemic income levels in 2022, while virtually all advanced economies are expected to.

As is so often the case, the people around the world who suffered the most got the least help. Those in poor countries were far less likely to get tested or treated for COVID than those in rich countries. The difference was most dramatic, though, with vaccines.

In January 2021, as COVID vaccines were rolling out, the director-general of the WHO began a board meeting with a grim assessment. "More than 39 million doses of vaccine have now been administered in at least 49 higher-income countries," said Dr. Tedros Adhanom Ghebreyesus. "Just 25 doses have been given in one lowest-income country. Not 25 million; not 25 thousand; just 25."

By May of that year, the gap that Tedros had warned of had become front-page news. "The Pandemic Has Split in Two" ran a headline in *The New York Times.* "Zero deaths in some cities. Thousands in others. The pandemic's fault lines continue to widen as vaccines flow toward rich countries." One WHO official denounced the inequity as "a moral outrage."

There were countless examples. By the end of March 2021, 18 percent of Americans had been fully vaccinated, while 0.67 percent of Indians and 0.44 percent of South Africans had been. By the end of July, the numbers had shot up to 50 percent in the U.S., but to just 7 percent in India and less than 6 percent in South Africa. Worst of all, people in wealthy countries who were at low risk of becoming severely sick were being vaccinated before people in poorer countries who were at a much higher risk.

To many observers, these facts were both infuriating and shocking. How could the world have billions of doses of a lifesaving vaccine and yet distribute them so unequally? Protesters marched, and politicians gave heartfelt speeches and pledged to donate doses.

Inside the world of global health, though, among people who work in the field, the reaction was different. Of course they were angry about the injustices of COVID. But they knew that COVID did not happen in a vacuum. Far from being the only inequity in global health, COVID wasn't even the *worst* inequity in global health.

Consider that by the end of 2021, COVID had caused more than 17 million excess deaths. It is impossible not to be horrified by this number. But compare it with the deaths in developing countries over the past decade:* 24 million women and babies died before, during, or shortly after childbirth. Intestinal diseases killed 19 million people. HIV killed nearly 11 million people, and malaria more than 7 million, most of them children and pregnant women. And this is just in the past ten years—these diseases have been killing people for far longer than that, and they won't go away when the pandemic does. They strike year after year, and unlike COVID, they are not at the top of the world's agenda.

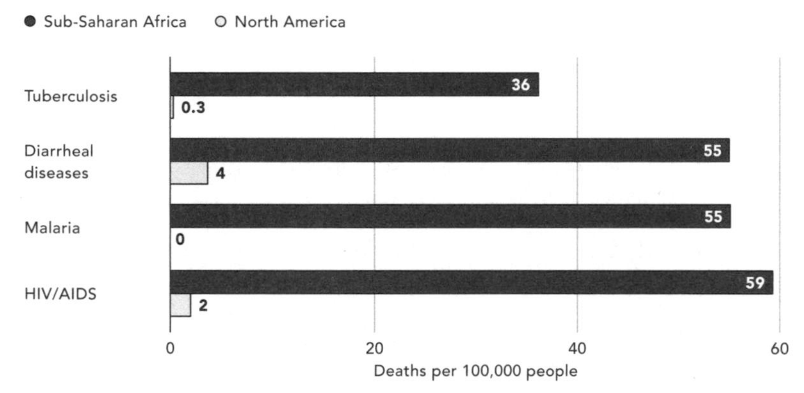

The health gap. Many people in sub-Saharan Africa die of diseases that rarely kill people in North America. (IHME)

The vast majority of people who die from these conditions live in low- and middle-income countries. Where you live and how much money you have determines—to a great extent—your chance of dying young or growing up to become a thriving adult.

* From 2010 to 2019, the most recent year for which data is available as this book goes to press.

Some of these diseases exist primarily in low-income tropical countries, which is why they are often ignored by much of the world. During the past decade, malaria killed 4 million children in sub-Saharan Africa but fewer than 100 people in the United States.

A child born in Nigeria is about twenty-eight times more likely to die before her fifth birthday than a child born in the United States.

A child born in the U.S. today can expect to live for seventy-nine years, but one born in Sierra Leone can expect to live just sixty years.*

In other words, health inequities are not rare. I think many people in rich countries were shocked by the world's unequal response to COVID not because it was out of the ordinary but because health inequities are not visible to them the rest of the time. Through COVID—a condition the whole world was experiencing—everyone could see how unequal the resources are.

The point is not to depress you or to wag a finger at people who haven't dedicated their lives to global health. The point is that all these problems deserve more attention. The fact that most of the people who suffer from these diseases live in low- and middle-income countries doesn't make them any less awful.

My dad had a beautiful way of elevating the moral dimensions of this phenomenon. Years ago, in a speech to the United Methodist Conference, he put it like this: "People suffering from malaria are human beings. They are not national security assets. They are not markets for our exports. They are not allies in the war against terrorism. They are human beings who have infinite worth without any reference to us. They have mothers who love them and children who need them and friends who cherish them—and we ought to help them."

* Differences in health usually hold true within countries as well as between them. In the U.S., Black women are three times more likely to die in childbirth than white women.

I could not agree more. When Melinda and I started the Gates Foundation two decades ago, we decided that providing resources to reduce and eventually get rid of this inequity would be our biggest focus.

The moral arguments do not fully persuade the governments of most rich countries to spend enough money to reduce or eliminate diseases that aren't killing their own people. Fortunately, there are also practical arguments that make the case even stronger, including the idea that better health makes the world more stable and improves international relations. I've been making that case for years, and now, in the COVID era, we have the benefit that investments in new medicines and health systems will help us stop pandemics before they engulf the world.

Virtually everything we should do to fight infectious diseases like malaria is also broadly helpful for future pandemics, and vice versa. It's not a binary choice in which we must decide whether to put money into pandemic prevention or infectious-disease programs. It is just the opposite. Not only *can* we do both, we *should* do both, because they are mutually reinforcing.

Let's review the progress that the world has made in global health, and what made that progress possible. As bad as the disparities I've mentioned above are, they are smaller today than at any point in history; when it comes to basic measures of health, we're moving in the right direction. How this progress happened is a thrilling story, and it bears directly on the world's ability to prevent pandemics.

I could cite dozens of statistics to show you the extent to which disparities in health have shrunk over the years. But I'm going to narrow it down to just one: child mortality.

From a clinical point of view, there's a good reason for using child mortality as a measuring stick for the health of the world. Improving child survival requires some interventions such as maternity care for their mothers, childhood vaccines, better education for women, and better diets. When more children survive, it's an indication that a country is getting better at doing these things.

But there's also another reason I use this statistic: When you look at health through the lens of child mortality, you can't help realizing just how high the stakes are. It is gut-wrenching to think about the death of a child. As a parent, I can't imagine anything worse, and I would give my life to protect my kids. For every child saved, there is a family that does not have to live through the worst anguish imaginable.

So let's look at how the world is doing on this fundamental measure of the human condition.

In 1960, nearly 19 percent of children died before their fifth birthday. Think about that for a moment: *Almost one in every five children on earth did not live to see the age of five.* And the disparity was enormous: In North America, the rate was 3 percent, while it was 21 percent in Asia and 27 percent in Africa. If you lived in Africa and had four children, you probably had to bury one of them.

Thirty years later, in 1990, the worldwide child mortality rate

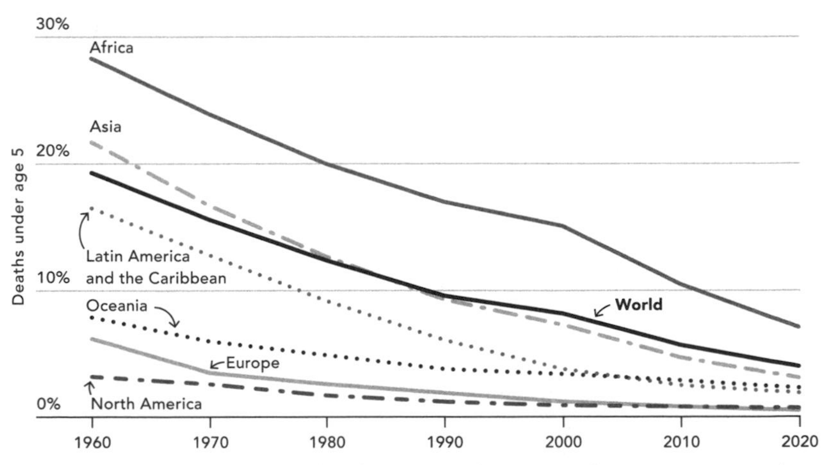

More children are surviving today than at any other point in history. In 1960, nearly 20 percent of children born did not reach their fifth birthday. Today that number is below 5 percent. (UN)

had dropped by half, to just under 10 percent. In Asia it was under 9 percent. Africa saw a similar but less dramatic improvement.

Now skip ahead another three decades to 2019, the latest year for which data is available. That year, less than 4 percent of the world's children died before age five. In Africa, though, nearly twice that many died.

I know this is a lot of numbers. To keep it simple, you can think of it as 20-10-5. In 1960, 20 percent of the world's children died. In 1990, it was 10 percent. Today it's less than 5 percent. The world keeps cutting the child mortality rate in half every thirty years, and we're on track to do it again well before 2050.

This is one of the great stories in human history, one that every high school student should know by heart. If you remember only one fact that illustrates the trajectory of human health over the past half century or so, it should be 20-10-5.

And yet 5 percent is still unbearably high. It amounts to more than 5 million children a year. Taken in isolation, preventing 5 million deaths sounds like an impossible task, but when you see the number in context, when you understand how far the world has come, it becomes a challenge and an inspiration to do even better. At least it did for me. It's the main focus of my work since I transitioned to working full-time at the Gates Foundation.

Over the years, I've given enough speeches on 20-10-5 and seen enough tweets and Facebook comments to know the question that inevitably comes next: Won't saving all these children lead to overpopulation?

It's a natural concern. It just seems like common sense that if more children survive, the global population will increase faster. In fact, I used to worry about this problem myself.

But I was wrong. The answer is, emphatically and without a doubt, no—lower rates of child mortality do not lead to overpopulation.

The best explanation of why this is true was given by my friend Hans Rosling. I first became aware of Hans when he gave an unforgettable TED talk in 2006 called "The Best Stats You've Ever Seen."* Hans had spent decades working in public health, with a focus on poor countries, and he used his talk to share some surprising facts about how health was improving around the world.

Eventually I got to meet Hans and spend a lot of time with him. I admired the clever, creative ways he showed people that the countries with the highest child mortality rates—places like Somalia, Chad, Central African Republic, Sierra Leone, Nigeria, and Mali—are also the countries where women have the most children.

When the child mortality rate drops, so does the size of the average family. It happened in France in the 1700s, Germany in the late 1800s, and in Southeast Asia and Latin America in the second half of the twentieth century.

There are various reasons that explain why this is the case; one factor is that—especially in places where there's no pension system or other support for the elderly—many parents feel that they need to have enough children so that they'll have someone to take care of them when they're older. If the odds are high that some of their children won't survive to adulthood, they'll make the perfectly rational decision to have more kids.

The drop in family size has led to a remarkable phenomenon: The world recently passed what Hans called "peak child"—that is, the number of children under five hit its maximum and is going down.† The benefit? As the United Nations Population Fund explains on its website, "Smaller numbers of children per household generally lead to larger investments per child, more freedom for

* It's at www.ted.com. I promise that you won't regret watching it.

† The global population will continue to increase for a while as the women born during "peak child" grow up and enter their reproductive years.

women to enter the formal workforce, and more household savings for old age. When this happens, the national economic payoff can be substantial."

So: Health is improving nearly everywhere, with big dividends for human welfare. The global health gap is still large, but it is narrowing.

As dramatic as this story is, it's only the background for what we need to know now. What caused these changes? And how can accelerating them also help prevent pandemics?

Trying to explain a decades-long global phenomenon involving billions of people is a risky endeavor. Entire books have been written on single aspects of the decline in child mortality and the march toward global health equity, and I'm covering the subject in just one chapter. I'm going to focus on the factors that are most directly relevant to the problem of preventing pandemics, with the understanding that I'm leaving out a host of others, including agricultural yields, global trade, economic growth, and the spread of human rights and democracy.

It is no accident that many of the tools used against COVID have their roots in global health. In fact, at virtually every step in the COVID response, there is an essential tool or system or team that exists only because the world has invested in improving health for the poor. The fingerprints of global health work are all over the COVID response.

Here's a list—and just a partial list at that—of ways that the two overlap.

Understanding the virus

Early in the pandemic, scientists needed to know what they were dealing with. To find out, they turned to genetic sequencing, the technology that sped up vaccine development (by quickly revealing the genetic code of the COVID virus) and made it possible to detect and monitor variants as they spread around the world.

It's not surprising that the first COVID variants were discovered somewhere other than the United States. The U.S. was slow to gather virus samples and get them sequenced; the lab capacity to do this existed—it just wasn't used. In the first year of the pandemic, compared with many countries, the U.S. was flying blind.

Fortunately, several countries in Africa—especially South Africa and Nigeria—were better prepared, having spent years building a robust network of sequencing labs. The original intent was to help with diseases that affect the continent disproportionately, but when COVID came along, these labs were ready to pivot; having been nurtured for years, they were able to produce more results, and faster, than their counterparts in the United States. South Africa's labs were the first to discover the Beta variant of COVID, as well as the subsequent Omicron variant.

Similarly, as I wrote in Chapter 3, computer modeling has helped us learn a lot about this pandemic, and it needs to be an even bigger part of our pandemic-prevention efforts. But the notion of using computer modeling to understand infectious diseases did not emerge suddenly with COVID.

The Institute for Health Metrics and Evaluation—whose computer models were widely cited by the White House and reporters during the pandemic—was created in 2007 to give the world insight into causes of death in poor countries. Imperial College London established its modeling center in 2008 with an eye toward assessing the risk of outbreaks and the effectiveness of different responses. That same year, I funded and recruited people for the Institute for Disease Modeling, which was designed to help researchers understand malaria better and to advise on the most effective paths to eradicating polio—and is now helping governments understand the impact of various COVID policies. The fact that these groups—and many others like them—turned out to be useful for COVID is a testament to the fact that investing in global health helps with pandemics too.

Getting lifesaving supplies

Another crucial early step, before vaccines were available, was to get preventive equipment (such as masks), oxygen, and other lifesaving equipment to people who needed it. This wasn't easy for anyone—even the United States struggled to acquire and deliver these things early on—and poor countries were in an even worse position. One of the organizations they could turn to for support was the Global Fund.

Created in 2002 to bolster the fight against HIV, TB, and malaria in low- and middle-income countries, the Global Fund has been a rousing success. It's now the world's largest nongovernment funder of this work. Today it makes sure that nearly 22 million people living with HIV/AIDS get the medicines they need to stay alive. Every year it distributes almost 190 million malaria-preventing bed nets—nets you hang over your bed at night to keep mosquitoes from biting while you sleep. In two decades, it has saved some 44 million lives. Years ago, I called the Global Fund the kindest thing that humans have ever done for one another. I still believe that today.

To do all this work, the Global Fund had to establish a solution for reaching people in need. It set up financing mechanisms so it could raise money and get it out the door quickly. It built systems to deliver medicines in some of the most remote places on the planet. It established networks of laboratories and set up supply chains.

When the Global Fund turned all these assets toward COVID, the results were impressive. It raised almost $4 billion for the COVID response in a single year, and it worked with more than 100 governments and more than a dozen programs that help multiple countries. Thanks to the fund, countries were able to buy COVID tests, oxygen, and medical supplies. They got protective equipment for their frontline health workers and stepped up their contact tracing efforts. Unfortunately, it wasn't all good news. Even though about a sixth of the additional money raised by the Global Fund supplemented its work on HIV, TB, and malaria, there were still big setbacks: TB deaths, for instance, rose in 2020 for the first time in more than a decade.

Making and testing new vaccines

When the effort to develop a COVID vaccine kicked into high gear, it relied heavily on work that had already been done for other diseases. For instance, mRNA technology had been in the pipeline for decades, with commercial funding to explore its potential as a cancer treatment and with government funding to develop it for fighting infectious diseases and bioterrorism.

Then, when it came time for human trials of vaccines—which is typically a long and expensive process, as you'll recall from Chapter 6—researchers turned to the HIV Vaccine Trials Network. As the name suggests, it was established to create an infrastructure that could speed up trials of HIV vaccines, a system that proved crucial for COVID vaccines. Although very few COVID vaccine trials took place in Africa, most of the ones that did relied on South Africa's strong clinical trial infrastructure, which had been built with funding for work on HIV vaccines. The first evidence of how effective COVID vaccines would be against a variant came from trials in South Africa.

Buying and delivering vaccines

Years ago, someone started a meme claiming that if I walked past a $100 bill lying on the sidewalk, it wouldn't be worth my time to pick it up. Although I've never had the opportunity to test this theory, I'm sure it's not true. I would absolutely pick up a $100 bill! First, I'd look around to see if I could find the person who dropped it, because someone is probably sad about losing that $100. And if I didn't see anyone, I'd take that money and send it where it could do the most good: Gavi, the vaccine organization I mentioned in Chapter 6.

Part of its mission is to help poor countries buy vaccines, but Gavi does much more than that. It also helps countries gather data to measure the effectiveness of their work and make improvements. It helps them build supply chains so that vaccines, syringes, and all the other necessary pieces reach the clinics where they're needed. It offers training for leaders in the health sector so they can manage their countries' vaccine programs more effectively and increase the public's demand for vaccines.

When the Gates Foundation helped create Gavi in 2001 with the goal of making vaccines available to all the world's children, we did not foresee the role it would play in combating a pandemic like COVID, but looking back, it now seems obvious: Gavi was a fantastic investment for saving children's lives that was also a fantastic investment for dealing with COVID. Having spent the better part of two decades helping poor countries improve their systems for delivering vaccines, Gavi had the skills and experience to help when global disaster struck.

Among other contributions, it's one of three partners running COVAX, the program designed to get COVID vaccines to people in developing countries. Although COVAX took longer than anyone hoped to reach its goals (for the reasons I explained in Chapter 6), it deserves credit for two important things. It delivered one billion doses of vaccines that had not even been available a year before, and it accomplished this feat faster than anything like it had ever been done before. (This was even more complex than it sounds: Although Gavi and UNICEF have built lots of infrastructure for delivering vaccines, their job is to immunize children and, in some cases, teenagers. They had to retool their systems to reach adults during COVID.)

It's not just global vaccination programs that are paying off for COVID. Countries that made a point of improving their own immunization efforts were also well positioned to respond. Let's look at one of them.

After it won independence from the U.K. in 1947, India undertook a massive campaign to eliminate smallpox—a project that required improving its health system, training vaccinators, buying cold chain equipment, reaching even the most remote parts of the country, and creating a surveillance network for vaccine-preventable diseases. It took decades, but it worked. India's last case of smallpox was reported in 1975.

Then, in the early 1980s, India turned to another problem: low rates of routine childhood immunizations. At the time, the percentage of children born in India who received these basic vaccines ran in the single digits. Building on the systems that had been put in place for smallpox, the government set out to raise immunization rates by an order of magnitude. It succeeded wildly: Vaccination rates soared, and case numbers plummeted. In 2000, for example, the country reported more than 38,000 cases of measles, and twenty years later it reported fewer than 6,000. Every year, India's immunization program gives primary doses to more than 27 million newborns, and booster doses to more than 100 million children between the ages of one and five years.

Building up a strong immunization program was a fantastic investment for India long before COVID came along, and when the virus did arrive, the investment paid off again. Because it had a system already in place, India rapidly established nearly 348,000 public centers and over 28,000 private centers to administer COVID vaccines—including many in the rugged mountain regions in the country's north and northeast. By mid-October 2021, the country had administered one billion doses of COVID vaccines. And building on existing systems, the government quickly set up a computer platform that allowed it to track the supplies of vaccines, record who had been vaccinated, and give people a digital certificate proving that they had been vaccinated.

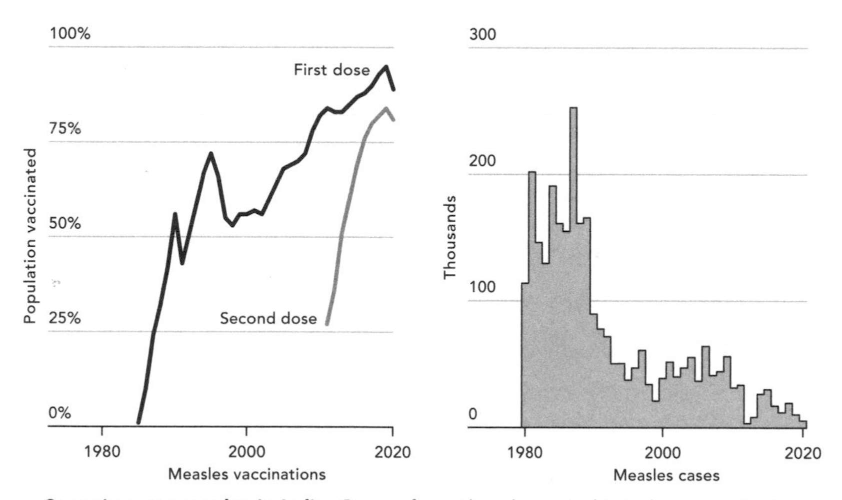

Stamping out measles in India. Cases of measles plummeted in India as vaccination rates soared. The first vaccine dose was introduced in India in the mid-1980s, and a second dose was added to the regimen years later. (WHO)

By mid-January 2022, one year after it started vaccinations, India had administered more than 1.6 billion doses, and more than 70 percent of adults there had received two doses. The government still had work to do, especially to vaccinate more people under the age of eighteen, but there is simply no way the country could have achieved as much as it did, as quickly as it did, without a well-run immunization program already in place.

Bringing it all together with logistics

Countries that had recently been running large polio campaigns—as both Pakistan and India had done—had another advantage: their national and regional emergency operations centers. (You may remember these nerve centers of public health campaigns from Chapter 2.) When COVID hit, these EOCs were a natural model for coordinating COVID-related activities.

In Pakistan, for example, health officials paused polio vaccination campaigns early in 2020 because of the transmission risk posed by vaccinators moving from one community to another. In March, though, they saw an opportunity: They could set up an emergency operations center for COVID modeled on the one for polio.

Within a few weeks, more than 6,000 health workers who had been trained to watch for signs of polio were taught about COVID symptoms as well. A call center that had been set up to take reports of possible polio cases was repurposed for COVID; anyone in the country could call a toll-free number to get reliable information from a trained professional. Staff from the polio EOC moved over to the COVID center to log case numbers, coordinate contact tracing, and share this information throughout the government—all functions that had been built up during the polio campaign. The maps, charts, and statistics pasted all over the walls now counted COVID cases.

And thanks to major investments in Pakistan's health system, the government there was ready to roll out COVID vaccines as soon as they became available. By late summer 2021, the country was vaccinating roughly a million people a day, a much higher share of its population than most other lower-middle-income countries, and by the end of 2021 it had doubled the rate to 2 million people a day.

This brings me to a criticism that I've been hearing for years. Trying to eradicate a disease is what people in the field call a vertical approach—that is, it goes deep on trying to end one disease. By contrast, a horizontal approach is one that can drive progress across many different problems at once. If you strengthen health systems, for example, you can expect to see improvements in malaria, child mortality, maternal health, and so on.

The criticism is that vertical efforts come at the cost of horizontal ones, and that those horizontal efforts are, by their nature, the more effective way to save and improve people's lives with limited money and effort.

I don't agree with this criticism. The way polio campaigns have shifted to help with COVID shows that horizontal and vertical capabilities are not a zero-sum game. And COVID is not the only example: During the 2014 Ebola outbreak in West Africa, polio workers in Nigeria were able to step in and help with the Ebola response. Without them, the country's nearly 180 million citizens would have been at far greater risk—and in fact, in countries without the polio-eradication infrastructure, the outbreak was much worse.

Strengthening one muscle does not have to come at the expense of weakening another. As we build up the world's ability to detect and respond to outbreaks—the most dangerous of which will be respiratory diseases—the investments we make will benefit the entire health system. The opposite is also true: When health care workers are well trained and have the tools they need, and when everyone gets good care, health systems will be able to stop outbreaks before they spread far and wide.

In my foundation work, I often advocate for increasing health aid to developing countries. Most people don't follow this subject, and they're surprised when they learn how little funding is involved.

If you added up all the money from governments, foundations, and other donors that helps low- and middle-income countries improve the health of their people, what would it amount to? We're counting everything: money for COVID, malaria, HIV/AIDS, child and maternal health, mental health, obesity, cancer, smoking cessation, and so on.

In 2019, the answer would've been $40 billion a year—that was the annual total of what's called development assistance for health. In 2020, when rich governments stepped up generously to deal with COVID, the answer was $55 billion. (As I write this, the number for 2021 isn't available yet, but I anticipate that it will be about the same.)

Whether $55 billion a year strikes you as a lot of money for global health depends on the context. It amounts to roughly .005 percent of the world's annual economic output. People spend almost that much on perfume every year.

Of that $55 billion annual figure, the United States contributes about $7.9 billion a year—more than any other country. That's less than 0.2 percent of the federal government's budget.

If you're a citizen in a donor country, you can feel good about the impact of this spending. The bang for your buck is tremendous.

Remember the 20-10-5 story from a few pages ago? That's what this money is paying for. This chart shows the dramatic decline in deaths of children under the age of five since 1990:

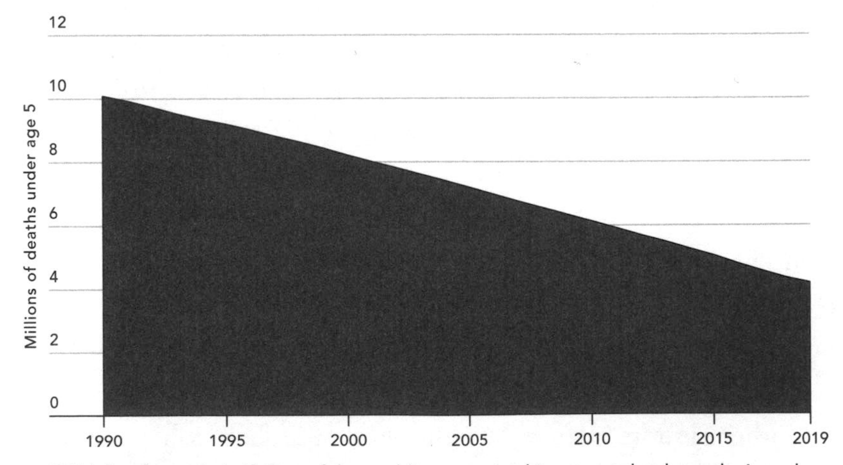

Child deaths cut in half. One of the world's greatest achievements has been the incredible progress in reducing child deaths. Here you can see the significant decline in deaths from infectious, nutritional, and neonatal diseases. (IHME)

And this chart shows you the world's progress on the worst causes of child mortality over the past thirty years:

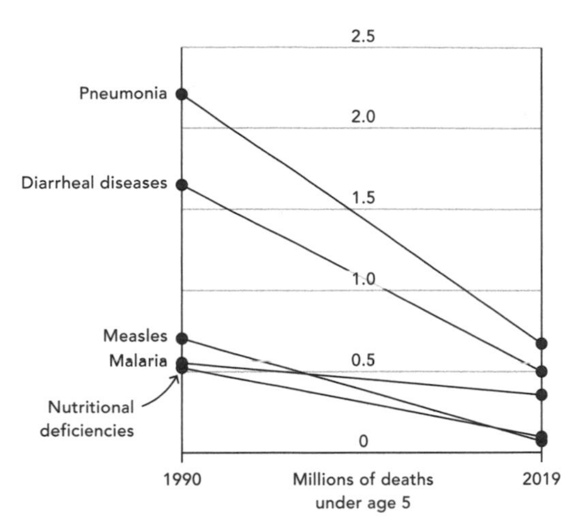

Going after preventable diseases. Investments in programs such as Gavi, the Global Fund, and the U.S. President's Malaria Initiative get a lot of credit for the massive decline in child deaths. (IHME)

See the big declines in deaths from diarrhea and pneumonia? Gavi gets much of the credit for that progress. See how malaria deaths also dropped? That's the Global Fund, along with government programs such as the U.S. President's Malaria Initiative.

This is historic progress on a global scale that translates into many millions of families who haven't had to bury a child. And, as we have now discovered, these efforts will provide another benefit: They will help prevent pandemics too.

CHAPTER 9

MAKE—AND FUND—A PLAN FOR PREVENTING PANDEMICS

One of the many lessons of COVID is that we all need to be cautious about predicting the course of a disease. This virus has defied expectations and surprised the scientific community many times, a fact that should be on the mind of anyone who is trying to look ahead—as I am doing with this chapter, written at the end of January 2022.

Based on what they know about the disease and its variants, many scientists now believe that by the summer of 2022 the world will be moving out of the acute phase of the pandemic. The number of deaths will be going down globally, thanks to the protection conferred by vaccines and by the natural immunity you get once you've had the virus. Countries with low COVID rates and high rates of other infectious diseases, such as malaria and HIV, may rightly decide to turn their attention back to these ongoing threats.

But even if that is how things turn out—and I hope it is—the work won't be over, because in all likelihood, COVID will become an endemic disease. People in low- and middle-income countries will still need better access to testing and treatments. Scientists will also need to study two key questions that will affect how the world lives with COVID. First: Which factors affect how immune you are to it? The more we understand these immune determinants, the better chance we have of keeping fatality rates low. The second question is: What is the impact of long COVID? Learning more about this syndrome (which I discussed briefly in Chapter 5) will help doctors treat people who have it and give public health officials insights into the burden it causes around the world.

Unfortunately, it is also possible that, whenever you read this chapter, we won't be out of the woods yet. A more dangerous variant may emerge—one that spreads more easily, causes more severe symptoms, or evades immunity better than previous variants. If vaccines and natural immunity don't prevent high death rates from such a variant, the world will have a major problem.

That's why national governments, academic researchers, and the private sector need to keep pushing for new or improved tools that will guard against the worst effects of COVID if the threat evolves. Governments will need to protect their people, using strategies that account for the fact that different places have different COVID profiles. The ability of new waves of COVID to spread through a population depends a lot on how many people have been vaccinated, infected, both, or neither. Health officials will need to adapt their strategies based on what the data says will be most effective in the areas where they're working.

To inform these efforts, governments need to push for better information on the incidence of COVID. Often, especially in developing countries, COVID data comes from limited clinical testing and from outdated information gathered in single surveys conducted in particular populations, such as health care workers or blood donors. With the help of ongoing disease surveillance, countries can gain crucial insights into subjects like the most effective ways to use nonpharmaceutical interventions while accelerating economic recovery.

With any luck, we will move to dealing with COVID as an endemic disease, the way we deal with seasonal flu. Meanwhile, whether COVID subsides or comes roaring back, we also need to work on a separate, longer-term goal: preventing the next pandemic.

For decades, people told the world to get ready for a pandemic, but hardly anyone made it a top priority. Then COVID struck, and stopping it became the most important thing on the global agenda. What I worry about now is that when COVID does subside, the world's attention will turn to other problems, and pandemic prevention will once again get moved to the back burner—or taken off the stove entirely. We need to take action now, while all of us still remember how awful this pandemic was and feel the urgency of never allowing another one to arise.

At the same time, experience can be misleading. We should not assume that the next pandemic threat will look exactly like COVID. It may not be so much worse for the elderly than it is for young people, or it may also spread by lingering on surfaces or through human feces. It may be more infectious, passing more easily from one person to another. Or it may be deadlier. Worst of all, it could be both deadlier *and* more infectious.

And it might be designed by humans. Although the world's plan should largely be focused on protecting us from natural pathogens, governments should also get serious about working together to prepare for a bioterror attack. As I argued in Chapter 7, most of this work involves steps we should be taking anyway, such as improving disease surveillance and preparing to make treatments and vaccines quickly. But defense officials need to be involved alongside health experts in designing policies, shaping the research agenda, and setting up disease simulations in which the pathogen is intended to kill millions or even billions of people.

Regardless of how the next big outbreak arises, the key is to have better plans than we do today and tools that can be deployed quickly. Fortunately, we have good systems in place for developing those tools. Governments in the U.S., Europe, and China fund early-stage experimental research and support product development work. India, Indonesia, and other emerging countries are taking steps in that direction too. Biotech and pharmaceutical companies have large budgets for taking ideas out of the lab and getting them to market.

What's lacking in most countries is a concrete plan—a national approach to research that funds the best scientific ideas. It needs to be clear who is driving the pandemic-related agenda, monitoring the progress on that agenda, testing ideas, implementing the most successful ones, and making sure they're turned into products that can be manufactured at mass volumes quickly. Without a plan in place, when the next major outbreak happens, the government's actions will be reactive, and too late. We'll have to try to figure out a plan after the pandemic is already spreading. That's no way to protect people.

Compare this situation to how governments approach national defense: They know exactly who's responsible for assessing threats, developing new capabilities, and practicing their deployment. We need outbreak strategies that are as clear, rigorous, and thorough as the world's best military strategy.

And don't forget that there is another big upside to all this work in addition to preventing pandemics: We can also eradicate entire families of respiratory viruses, including coronaviruses and influenzas—viruses that cause immense suffering and hardship. The impact on human lives and economies around the world would be phenomenal.

I see four priorities for a global plan to eradicate respiratory diseases and prevent pandemics. After I lay out each one, I'll walk through the funding that will be required.

1. Make and deliver better tools.

My work in technology and philanthropy stems from a simple idea: Innovation can improve lives and solve important problems, whether it's making education more accessible to more people or reducing childhood deaths. In the past few decades alone, advances in biology and medicine have unlocked new ways to treat and prevent disease.

But innovation does not simply happen on its own. As the story of mRNA vaccines shows, ideas must be nurtured and researched, sometimes for decades, before they produce anything of practical value. That's why step 1 in any pandemic-prevention plan should be to keep investing in better vaccines, therapeutics, and diagnostics.

Although mRNA vaccines are extremely promising, public and private researchers alike should pursue other approaches, such as the adjuvanted-protein vaccines that I described in Chapter 6, since they might protect people for longer, reduce the number of breakthrough infections, or attack parts of the virus that aren't likely to change in future variants. Ultimately, our goal should be to develop novel vaccines that fully protect against entire families of viruses, particularly respiratory viruses—that's the key to eradicating flu and coronaviruses. All of the players involved in researching and developing vaccines—including government and philanthropic funders, academic researchers, biotech companies, and pharmaceutical developers and manufacturers—need to help identify the best early-stage ideas and advance them all the way to a product.

In addition to vaccines, we should also be pursuing infection-blocking drugs, which people can give themselves to get immediate protection from a respiratory pathogen. Governments should create incentives to develop and use this approach—including, once blockers are available, federal reimbursements to doctors who prescribe them to their patients, just as they get reimbursed for other medicines and vaccines.

We also need to improve our ability to test and approve new products—as you saw in Chapters 5 and 6, it's a time-consuming process. A few efforts, such as the RECOVERY trial in the U.K., set up protocols in advance and built infrastructure that made it much easier to get started once COVID hit. We should build on those models, improving the ability to run trials around the world so we can quickly learn what works even when a new disease is present in just a few countries. Regulators need to agree ahead of time on how people will be enrolled and on the software tools that will enable people around the world to sign up as soon as the disease strikes. And by plugging diagnostic reports into the trial system, we can automatically suggest to doctors that their patients should join a large-scale trial.

We also need to be ready to make lots of doses fast. The world needs large-scale manufacturing capacity—enough to provide everyone on the planet with all the necessary doses of a new vaccine within six months of identifying a pathogen that could go global. During COVID, when countries that make lots of vaccines were hit hard by the pandemic, they restricted exports of their vaccines to make sure they had enough for their own people. The world's interest, though, is in vaccinating everyone, so we'll need to account for this complicating factor by investing in more manufacturing capacity and in innovations that make it easier to do technology transfers and second sourcing.

Manufacturers in China and India are experts at producing new tools in large numbers and can be part of the solution. Different countries can commit to provide a portion of the manufacturing capacity required. If China, India, the United States, and the European Union each agree to provide a quarter of the capacity in the near term and countries in Latin America and Africa keep developing their facilities, we will have a global solution.

Another crucial area for research is to make it easier to *deliver* vaccines by, for example, solving the cold chain problem. Micro-needle patches would do that while also making it less painful to get a vaccine and allowing people to vaccinate themselves. Measles vaccines that use micro-needle patches are in the pipeline, but a lot of work remains to make the approach cheap enough to use in large numbers.

Other promising ideas include vaccines delivered via nasal spray, ones that give protection for decades, ones that are given in a single shot without the need for follow-up doses, and combination vaccines that work against multiple pathogens (a combined flu-and-COVID shot, for example).

If making vaccines within a year was the surprise success story of COVID, the fact that it took so long to develop effective therapeutics was the surprise disappointment. Despite what I and others hoped early on, it took nearly two years to find effective antivirals for COVID—and in a pandemic, two years is an eternity. As we roll out the treatments that we now have, we should also build the systems that will allow us to develop and deliver treatments much faster in the future.

One essential step is to prepare a library of millions of antiviral compounds that are designed to attack common respiratory viruses, including drugs that work on a wide range of variants. If we have three or more of these compounds, we can combine them to reduce the chances that a drug-resistant variant will emerge. (This is done now with HIV treatments: Three antivirals are combined, limiting the spread of resistant viruses.) All researchers should have access to these libraries so they can see which compounds already exist and which areas of research will be the most fruitful. They should also be studying long COVID so they can understand what's driving it, how to help people who are suffering from it, and whether future pathogens may have similar long-lasting symptoms.

Another important step is to take advantage of advances in artificial intelligence and other software to develop antivirals and antibodies faster. Several companies are doing great work in this field. Essentially, you would build a 3D model of the pathogen that you want to target—it could even be one we've never seen before—as well as models of various drugs that you think might work against it. The computer would quickly run these models against one another, tell you which drugs look promising, figure out how to improve them, and, if necessary, even design new ones from scratch.

We should also expand the incentives for generic manufacturers to make antiviral treatments available sooner than they were for COVID. Advance orders on behalf of low- and middle-income countries do just that, by getting generic drugmakers to start manufacturing a new drug even while it's still going through regulatory approvals. (These advance orders eliminate the risk that the generic companies will lose money if the approvals don't come through.)

One final note about biomedical research. A great deal has been written about how and where COVID began. My own view is that the evidence is very strong that it jumped from an animal to a human and did not, as some people have argued, come from a research laboratory. (I know some well-informed people who don't think the evidence for this view is as solid as I do. The question may never be resolved to everyone's satisfaction.) But regardless of how COVID started, even the remote possibility of lab-related pathogen releases should inspire governments and scientists to redouble their efforts on lab safety, creating global standards and inspections for infectious-disease facilities. The last known smallpox death anywhere occurred in 1978, when a medical photographer at the University of Birmingham got the disease because of a leak in the building where she worked, which also housed a lab studying smallpox.

In addition to deploying better vaccines and treatments, we need to spark more innovation in diagnostics. Testing people for a disease should serve two purposes: letting them know very quickly whether they're infected so they can take action (including isolating themselves), and also informing public health officials so they know what's happening in the community. Some portion of positive tests should be gathered and sequenced so we can rapidly detect and understand emerging variants. The quick rollout of PCR tests and quarantine policies explains in large part why some countries, such as Australia, had dramatically fewer infections and excess deaths than others. Governments need to learn from these examples and figure out how they will ramp up testing very quickly—and give people an incentive to get tested by also offering treatment to anyone who tests positive and is at significant risk of severe disease.

Researchers should keep working on—and funders should keep supporting—high-throughput PCR tests, which have all the benefits of a typical PCR test but are also remarkably fast. They're also very cheap, they don't require the supplies of reagents that limited our diagnostic capacity during COVID, and they're easy to adapt to detect a new pathogen as soon as we've sequenced its genome.

We also need to support work on new types of tests that make it easier to collect samples and turn around results quickly. Low-cost diagnostics that resemble a pregnancy test—the ones known as lateral flow immunoassays—unlock the possibility of testing throughout entire communities. We can also deploy machines such as the LumiraDx, which I mentioned in Chapter 3: They can be used for a broad range of tests that already exist, and they can be adapted quickly for new tests too. And if in a future outbreak self-swabbing is an effective way to get a sample, as it is for COVID, we'll be able to use this technique to expand testing rapidly even in low-income countries.

2. Build the GERM team.

The group I envisioned in Chapter 2 will take years to fully assemble, which is why the work needs to start now. For the GERM team to become a reality, governments will need to provide resources and make sure it's well staffed. Many organizations can offer advice on the design of GERM, but its annual budget needs to be almost entirely paid for by the governments of wealthy countries and managed through the WHO as a global resource.

To make the most of the money and effort going into the GERM team, the world will also need to invest more in a complementary area: public health infrastructure. This isn't about doctors, nurses, and clinics—we'll come to those later in this chapter—but rather the epidemiologists and other specialists who do disease surveillance, manage the response to an outbreak, and help political leaders make informed decisions during a potential crisis.

Public health institutions don't get as much public attention or government funding as they deserve—not at the state level (including in the U.S.), national level, or global level with the WHO. That's not surprising, since their work is largely focused on preventing disease, and as public health experts like to say, no one thanks you for the disease they didn't get. Yet because of this inattention, many elements of public health departments need to be modernized, including the way they recruit and retain great people as well as the software they use. (In 2021, Microsoft worked with one U.S. state health department whose software was two decades old.) They're the foundation for a quick and effective response during an outbreak, and they need to be bolstered.

3. Improve disease surveillance.

After a lifetime of neglect by the public at large, disease surveillance is finally having its moment. The world has a lot of catching up to do.

One crucial step is to improve civil registration and vital statistics in the developing world. At a minimum, many low- and middle-income countries need stronger registries of births and deaths, information that would feed into national disease surveillance work, such as the effort in Mozambique that I described in Chapter 3. Then, building on that foundation, they should expand into genomic sequencing, autopsies that use minimally invasive tissue sampling, wastewater surveillance, and other practices. Ultimately, the goal for nearly every country is to be able to detect and respond to outbreaks within its borders—whether the disease is TB, malaria, or one we've never seen before.

In addition, the world's disparate disease surveillance systems need to be integrated so that public health officials can rapidly detect emerging and circulating respiratory viruses no matter where they emerge. These systems should use both active and passive approaches and make the data available in real time—because outdated data is not only unhelpful, it's often misleading too. As I've emphasized throughout this book, test results need to be tied into the public health system so that health officials can watch for outbreaks and understand endemic diseases better; the Seattle Flu Study is a good model to build on. And in countries like the United States, where testing can be extremely expensive, governments need to create incentives to make diagnostics cheaper and more accessible to everyone.

Finally, we need to expand our capacity to sequence the genomes of pathogens. The effort to do this in Africa has paid off—sequencing that was done on the continent alerted the world to at least two COVID variants—and this is the right time to double down on these investments by, for example, supporting projects like the Africa Pathogen Genomics Initiative, a network of labs across the continent that are able to share genomic data with one another. There's a similar network in India, and the model is being expanded to South Asia and Southeast Asia, but it should go even further. China too is home to a very effective sequencing industry, which needs to be part of the global system. Sequencing work has many benefits, even beyond preventing another pandemic—for instance, it will give governments new insights into the genetics of mosquitoes and malaria and the transmission of tuberculosis and HIV.

The field of genomics would also benefit from more investment in advances like the Oxford Nanopore sequencer and smartphone app that I mentioned in Chapter 3, which make it possible to sequence genomes in many more places. There should also be more research into how changes in a pathogen's genetic makeup affect the way it operates in the human body. Today we can map the mutations in different versions of a pathogen, but will a given mutation make the variant more transmissible? Will it lead to illnesses that are more severe? We don't know much about the answers to these questions, and it's a rich field for scientific inquiry.

4. Strengthen health systems.

When I first got involved with global health, I was focused like a laser on developing the kinds of new tools I've been describing. *Make a new rotavirus vaccine,* I thought, *and no kids will die of rotavirus.* Over time, though, I've seen how limitations in the health care delivery system—particularly the basic level, which is called the primary health care system—prevent vaccines and other new tools from reaching all the patients who need them.

A major part of the Gates Foundation's work has been to help improve these systems and ensure that they can reach all kids with new vaccines—investments that both save lives and lay the foundation for economic growth.* Once a country moves out of poverty and achieves middle-income status, its government will cover its own health care needs. Many countries have made this transition over the past few decades, and today, less than 14 percent of the global population lives in low-income countries that still need help financing basic health care.

* It's also important that scientists who are working on breakthroughs prioritize making them cheap and practical enough to work everywhere, not just in high-income countries. Delivery should be baked in from the beginning.

The pandemic devastated health systems around the world—the WHO estimates that by May 2021, more than 115,000 health care workers had died of COVID—but the need in low-income countries is especially acute. The fundamental challenge is that they don't have the funding, experts, or institutions they need to offer basic health services to all their people, let alone manage a major outbreak. And during the pandemic, the problem got worse, as many rich governments cut foreign aid or took money from work on other diseases and redirected it to COVID.

We need to reverse this trend. The models for wealthy governments are still Sweden and Norway, both of which give at least 0.7 percent of their GDP in aid for low- and middle-income countries, and much of that money is spent specifically on improving health. (I will return to the 0.7 percent target shortly.)

For their part, low- and middle-income countries should learn from the many good examples around the world. Sri Lanka, for instance, spent years building a strong primary health care system, which helped bring infant and maternal mortality down significantly, even when the country was still very poor.

As they rebuild, governments should focus on health spending that achieves many things at once. For example, hiring more health care workers gives you more people who can manage malaria cases, offer HIV testing and treatment, and do contact tracing for TB patients. And, armed with new digitally connected diagnostics—such as a handheld ultrasound device that can help assess the health of a fetus and detect viral pneumonia, TB, and breast cancer—these workers can form the backbone of a dynamic health system that gives public officials unprecedented insight into what's causing illness and death in their country.

But as COVID made obvious, low- and middle-income countries aren't the only ones that need to strengthen their health systems. Although there were a few exemplars who took early action, no country's response was perfect. So there are a few steps that countries at every income level should consider.

One is to focus more on primary health care. In many low-income countries—and the United States as well—most of the national health spending goes to expensive hospital care for people with advanced illness, while primary care goes underfunded. But studies have shown that putting more into primary care can actually drive down overall health costs: If high blood pressure is diagnosed early through the primary care system, the patient can get inexpensive medication and counseling and avoid the life-threatening and expensive consequences—heart attack, kidney failure, stroke—that will require an expensive visit to the hospital. It's been estimated that 80 percent of health problems can be handled effectively by a strong primary health care system.

Another crucial step is to decide, in advance of a crisis, who's responsible for doing what. Outbreak simulations like Crimson Contagion highlighted the possibility for chaos—remember the problem with the names of conference calls?—but hardly anything was done about it. Now we know the consequences of indecision.

During COVID, and especially in the early days, there was a lot of confusion in the U.S. about what the states could or should do, and what role the federal government would play. Similarly in Europe, there was some confusion over whether individual countries or the EU would be responsible for buying vaccines. In an emergency, the last thing you want is a lack of clarity that leaves people unsure about their own responsibilities.

Each country needs a pandemic prevention czar with the mandate to establish a plan and then execute it to contain an outbreak. That person's authority needs to include establishing rules for procuring and distributing essential supplies, as well as having access to data and modeling. The GERM team should play this role internationally.

Governments and donors also need a global forum where they can coordinate action with and on behalf of poor countries—for example, by agreeing ahead of time on how they'll unlock financing to purchase vaccines, tests, and other products, so they can avoid fundraising during a crisis. They should also agree in advance on the principles that will guide distributing these products, so new tools can reach the people who need them faster.

In the United States, the federal government is in the best position to drive large-scale development and production of vaccines, treatments, and personal protective equipment. But management of testing and hospital resources is inherently more local. And what about something like vaccine distribution? Although there will always be national and even global supply chains, the last mile for distribution is inherently local in nature. Japan did a good job of clarifying responsibilities at different levels and is a good model for others to study.

Each government's plan needs to account for the distribution of all the necessary tools, including masks, tests, treatments, and vaccines. This is not solely a problem in low- and middle-income countries; almost every government struggled to deliver vaccines during COVID. Building better data systems will make it easier to see where supplies are needed and to verify who's been vaccinated. Some countries, such as Israel, handled the verification process well during COVID, but others made a mess of it.

You can't improve the delivery of health care overnight, so the countries that make the effort in nonpandemic times will be much better off when things go wrong. If you have an established supply chain and workforce for educating people about Ebola transmission or delivering measles vaccines, then you have a playbook to start with and a team to read it. As Bill Foege once told me: "The best decisions are based on the best science, but the best results are based on the best management."

—

The world's richest countries have a proud history of leading the way on innovation. The U.S. government, for instance, supported the research that led to the creation of the microchip, unleashing a flurry of advances that made the digital revolution possible. Without those investments, Paul Allen and I would never have been able to even imagine a company like Microsoft, much less actually make it happen. Or take a more recent example: the groundbreaking work being done on zero-emissions energy sources at national labs across the country. If the world can manage to eliminate greenhouse gas emissions by 2050, as I think it can, the energy research supported by the U.S. and its peers will be one reason why.

When COVID struck, crucial advances in vaccines were made by academics and companies in the United Kingdom and Germany. Funding from high-income countries—particularly the United States (one area in which it led the world)—helped accelerate the innovations that would prove to be essential in fighting the disease. One part of the U.S. government supported the academic work on mRNA, another part supported the effort to translate the basic research into marketable products, and another part funded vaccine companies that were working with mRNA and other vaccine technologies when the pandemic struck.

Now governments must keep leading the way with new funding for the systems, tools, and teams that the world needs to prevent pandemics. As I wrote in Chapter 2, I believe the GERM team will require roughly $1 billion per year, funding that should be supplied by the governments of rich and some middle-income countries.

One of GERM's jobs will be to help identify the most promising new tools. I estimate that, over the next decade, all governments combined need to spend $15 billion to $20 billion per year to develop the necessary vaccines, infection-blocking drugs, treatments, and diagnostics—a level of spending that we can reach if the U.S. increases its health research spending by 25 percent, or roughly $10 billion, and if the rest of the world matches that increase dollar for dollar. Of course, $10 billion is a lot of money in absolute terms, but it is just over one percent of the U.S. defense budget, and it's a drop in the bucket compared to the trillions lost during COVID.

To make the most of these new tools and the GERM team, we'll need to take on the fundamental work of strengthening health systems (the clinics, hospitals, and health care workers who see patients) as well as public health institutions (the epidemiologists and other health officials who watch for and respond to outbreaks). Because the world has a long history of underinvesting in both areas, there's a lot of catching up to do: Getting high- and middle-income countries ready to prevent pandemics will cost at least $30 billion per year—that's the total for all of these countries combined.

This work needs to be done in low-income countries too, which is why it's so important for all wealthy countries to be as generous as Norway, Sweden, and the other governments that invest at least 0.7 percent of their GDP in development aid. If every country reaches that level, it will generate tens of billions of dollars in new money for strong health systems—money that, as I argued in Chapter 8, can be used to save children's lives *and* stop pandemics before they start.

The idea that rich countries should dedicate at least 0.7 percent of their GDP to aid has a long history, dating from at least the late 1960s. In 2005, the European Union committed to reaching the goal by 2015, and although many of the world's governments are quite generous, only a few have met their pledge. Today, when COVID makes it undeniable that health in one part of the world matters to every other part of the world, there has never been a better moment for rich governments to rededicate themselves to this goal. Investing in the health and development of low-income countries is good for the entire world: It makes everyone safer and more secure, it's a foundation for growth that helps people and countries get out of poverty, and it's the right thing to do.

More funding is necessary, but it's not sufficient. Another key contribution will be to ease the path to approval for products without sacrificing safety. As the scientists behind the Seattle Flu Study and SCAN saw firsthand, it's too difficult and time-consuming to put breakthrough ideas to work, especially during an emergency, when every hour counts.

Meanwhile, leaders of low- and middle-income countries should make it a priority to detect and stop outbreaks, seeking outside technical support and funding when they'll be helpful. And by joining projects like the global system for sharing health data, they can give themselves and the world more insight into what's happening on the ground in every region.

As the organization responsible for coordinating the GERM team, the WHO can help by prioritizing GERM's primary mission: detecting outbreaks and sounding the alarm. GERM has a secondary mission—to help reduce the burden of infectious diseases, including malaria, measles, and others—that will help save hundreds of thousands of lives and allow the team members to keep their skills sharp when they're not actively fighting an outbreak.

The WHO is the only organization that can strengthen the requirements for governments to be more open about potential outbreaks within their borders. Member countries of the WHO can also hold one another accountable for doing so—while recognizing that there are incentives to do just the opposite. If sharing news of a possible outbreak means that a country will be singled out for travel restrictions, it could hurt the local economy—a strong reason not to do it. But the global community has an interest in getting this information, and governments around the world have committed to sharing it as part of the International Health Regulations. The

WHO should work with its member states to strengthen these regulations and their implementation. As we've learned from COVID, the countries that shared information and acted quickly paid a short-term price—there's no question that lockdowns and travel bans are painful even when they're appropriate—but they kept the damage from being as bad as it could have been, for their own people and the rest of the world too.

Other groups also have important roles to play. Pharmaceutical and biotechnology companies should commit to more tiered pricing and second-source deals to make sure that even their most advanced products are available to people in developing countries. Technology companies should help develop new digital tools, such as ways to make it easier and cheaper to collect samples for diagnostic tests, or software that monitors the internet for signs of an outbreak.

More broadly, foundations and other nonprofits should help governments bolster their systems for public health and primary health care. The public sector will always pick up most of the check and do the heavy lifting on implementation, but nonprofits can test out new ideas and identify the ones that work best. Foundations should also support research on better tools that can be used for both today's infectious diseases and future pandemic threats. And because other global problems don't pause for a pandemic, philanthropy also needs to keep supporting efforts to avoid a climate disaster, help low-income farmers grow more food, and improve education around the world.

When I started telling friends that I was working on a book about pandemics, I could see that they were a little surprised. Many of them had been nice enough to read the book on climate change I published in 2021, and although they were too polite to say it, they were clearly thinking, "How many more of these books are you going to write where you're telling us about some big problem and a plan to solve it? We have to do climate. Now we're doing pandemics and health. What else is there?"

The answer is that these are the two major problems I think we need to put more resources into. Climate change and pandemics—including the possibility of an attack by bioterrorists—are the most likely existential threats for humans. Fortunately, there are opportunities to make major progress on both of them in the next decade.

For climate change, if we spend the next ten years developing green technologies, setting up the right financial incentives, and getting the right public policies in place, we can be on a path to net zero emissions of greenhouse gases by 2050. The news is even better for pandemics: Over the next decade, if governments expand their investments in research and adopt evidence-based policies, we can develop most of the tools we need to keep an outbreak from turning into a disaster. The amount of money required for pandemic preparedness is far less than what it will take to avoid a climate disaster.

This cause may seem remote. It can be hard to feel that you have any ability to affect the course of a pandemic. A mysterious new disease is frightening, and it can also be frustrating, because it seems that we have so little power to do anything about it.

But there are things that each of us can do. Elect leaders who will take pandemics seriously and make good, science-based decisions when the time comes. Follow their advice about masking up, staying home, and keeping your distance when you're out. Get vaccinated when you can. And avoid the misinformation and disinformation that flood social media: Get your information regarding public health practices from reliable sources, such as the WHO, the CDC in the United States, and its equivalent in other countries.

Most of all, don't let the world forget how awful COVID was. Do whatever you can to keep pandemics on the agenda—locally, nationally, internationally—so we can break the cycle of panic and neglect that makes them the most important thing in the world for a time, until we forget about them and go back to our daily lives.

We're all eager to return to the way things were before, but there is one thing we cannot afford to go back to—our complacency about pandemics.

We don't need to surrender to living in perpetual fear of another global catastrophe. But we do need to remain aware of the possibility and be willing to do something about it. The fact that we now understand the threat like never before should inspire the world to take action—to invest billions now so we won't lose millions of lives and trillions of dollars in the future. This is an opportunity to learn from our mistakes and ensure that no one ever has to live through another disaster like COVID. But we can be even more ambitious than that: We can work toward a world where everyone has a chance at a healthy and productive life. The opposite of complacency isn't fear. It's action.

AFTERWORD

HOW COVID CHANGED THE COURSE OF OUR DIGITAL FUTURE

While I was writing this book, I spent a lot of time thinking about how the COVID pandemic accelerated innovation on the infectious-disease front. But it also ushered in a new era of rapid change that goes way beyond health innovations.

In March 2020, when most of the world adopted strict lockdown rules, many people were forced to find ways to replicate in-person experiences from the safety of their homes. In places like the United States,* we turned to digital tools like videoconferencing and online grocery shopping to fill the gaps, using them in new, creative ways. (I remember thinking that the concept of a virtual birthday party seemed so strange back in the early days of the pandemic.)

I think we'll look back at March 2020 as an inflection point when digitization began to accelerate rapidly. Although the world has been becoming more and more digitized for decades, the process has been relatively gradual. In the U.S., for example, smartphones seemed to become ubiquitous overnight, but it took ten years to go from 35 percent of Americans owning one to today's level of 85 percent.

* The pandemic has accelerated digitization all over the world in different ways, but I'm going to focus on high-income countries, where the pace of change has been the most dramatic.

March 2020, on the other hand, was an unprecedented moment when digital adoption took a huge leap forward in many areas. The changes weren't limited to any demographic or specific technology. Teachers and students turned to online platforms to keep learning. Office workers began holding brainstorming sessions over Zoom or Teams before hosting virtual trivia nights with their friends in the evening. Grandparents signed up for Twitch accounts to watch their grandkids' wedding ceremonies. And just about everyone started shopping online a lot more, causing e-commerce sales in the United States to jump by 32 percent in 2020 compared with the year before.

The pandemic forced us to rethink what is acceptable for many activities. Digital alternatives that were once viewed as inferior were suddenly seen as preferable. Before March 2020, if a salesperson offered to make a pitch via video conference, many customers would have taken that as a sign that he or she wasn't serious about winning their business.

Pre-pandemic, I wouldn't have dreamed of asking political leaders to spend thirty minutes on a video call discussing how to improve their primary health care system, because it would have been seen as less respectful than meeting in person. Now when I suggest a video call, they understand how effective it can be and set time aside to meet virtually. Once people learn the digital approach, they generally stick to it.

In the early days of the pandemic, many technologies were merely "good enough." We were using them in ways that weren't exactly their intended purposes, and the results were sometimes bumpy. Over the past couple of years—as it's become clear that the need for these digital tools is here to stay—we've seen huge improvements in quality and features. These advancements will only continue in the years ahead as both the hardware and software get better.

We're just at the beginning of this new era of digitization. The more we use digital tools, the more feedback we get about how to make them better—and the more creative we'll be about how we can use them to improve our lives.

The first book I wrote, *The Road Ahead,* was all about how I thought personal computers and the internet would shape the future. It came out in 1995, and although I didn't nail all of my predictions (I thought digital agents would be almost as good as human assistants by now), I also got some key things right (we now have video on demand and computers that fit in our pockets).

This is a very different book. But just like *The Road Ahead,* it is fundamentally about how innovation can solve big problems. And I wanted to share some thoughts about how technology will change our lives even faster due to the need to rethink our approaches during the pandemic.

One of my favorite authors, Vaclav Smil, has this riff he uses in several books. He tells you about a young woman who wakes up and drinks a mug of instant coffee before taking a subway to work. When she gets to the office, she takes an elevator to the tenth floor and stops to grab a Coca-Cola from the vending machine on the way to her desk. The plot twist is that the situation he's describing takes place in the 1880s, not the modern era.

When I first heard his riff years ago, I was struck by how familiar the scene Smil described was. But when I read it again during the pandemic, it felt for the first time as if he was describing the past (although not the part about drinking a Coke in the middle of the workday!).

Of all the areas that are forever changed by the pandemic, I suspect that office work will see the most dramatic shift. The pandemic disrupted work in virtually every industry, but office workers were in the best position to take advantage of digital tools. The situation Smil describes—where you commute somewhere every day and work from a desk in an office—sounds increasingly like a relic of the past, even though it was the norm for more than a century.

As I write this in early 2022, many companies and workers are still figuring out what their "new normal" looks like. Some have already returned to full in-person work. Others have committed to being entirely remote. Most are somewhere in between, still trying to figure out what works best.

I'm excited about the potential for experimentation. Expectations around traditional work have been upended. I see lots of opportunity to rethink things and find out what is effective and what isn't. Although most companies will likely opt for a hybrid approach in which people come into the office part of the week, there's a good deal of flexibility around what exactly that looks like. What days do you want everyone to be in the office for meetings? Do you let people work remotely on Mondays and Fridays, or do you let them stay home in the middle of the week? In order to minimize commuter traffic, it would be best if every company in an area didn't pick the same days.

One prediction I made in *The Road Ahead* was that digitization would create more choices about where to live and lead to many people moving farther out of cities. It looked as though this wasn't going to pan out—until the pandemic hit. Now I am doubling down on that prediction. Some companies will decide that time in the office is required only one week a month. This will allow employees to live farther away, since a long commute is easier to tolerate if you aren't doing it most days. Although we've seen some early signs of this transition, I think we'll see a lot more of it in the decade ahead as employers formalize remote-work policies.

If you decide that employees are required to be in the office less than 50 percent of the time, you can share your workspace with another company. Office space is a significant expense for businesses, which could be cut in half. If enough companies do this, the demand for expensive office space would be reduced.

I don't see any reason why companies need to make firm decisions right away. This is a great time to take an A/B testing approach. Maybe you have one team try one configuration while a different team tries another, so that you can compare the results and find the right balance for everyone. There will be tension between managers who tend to be more conservative about new approaches and employees who want more flexibility. Résumés in the future will likely include information about preferences for being able to work away from the office.

The pandemic has forced companies to rethink productivity in the workplace. The boundaries between once discrete areas—brainstorming, team meetings, quick conversations in the hallway—are collapsing. Structures that we thought were essential to office culture have begun to evolve, and the changes will only intensify in the years to come as businesses and employees settle into new permanent ways of working.

I think most people will be surprised by the pace of innovation over the next decade now that the software industry is focused on remote working scenarios. Many of the benefits of working in the same physical space—like running into people at the water cooler—can be re-created with the right user interface.

If you use a platform like Teams for work, you're already using a much more sophisticated product than you were in March 2020. Features like breakout rooms, live transcription, and alternate viewing options are now standard across most teleconferencing services. Users are just beginning to take advantage of the rich features available to them. For example, I often use the chat function in many of my virtual meetings to add comments and ask questions. When I meet in person now, I miss the ability to have this kind of high-bandwidth interaction without distracting the group.

Eventually, digital meetings will evolve beyond simply duplicating an in-person meeting. Live transcription will one day allow you to search for a topic across all meetings at your company. You might be able to have action items automatically added to your to-do list as they're mentioned, and analyze a meeting's video recording to learn how to make your time more productive.

One of the greatest drawbacks of online meetings is that video doesn't let you see who is looking where. A lot of the nonverbal exchanges get lost, eliminating a human element. Moving from squares and rectangles to other "seating" arrangements makes things a bit more natural, but it doesn't solve the loss of eye contact. This is about to change as we move participants into a 3D space. A number of companies—including Meta and Microsoft—have recently unveiled their visions for the "metaverse," a digital world that both replicates and enhances our physical reality. (The term was coined in 1992 by Neal Stephenson, one of my favorite modern science fiction authors.)

The idea is that you will use a 3D avatar—a digital representation of yourself—to meet with people in a virtual space that mimics the feeling of being together in real life. This feeling is often referred to as "presence," and a lot of tech companies have been working on capturing it since before the pandemic started. When done well, presence can not only replicate the experience of an in-person meeting but enhance it: Picture a meeting where engineers at a car company who live on three different continents pull apart a 3D model of a new vehicle's engine to make improvements.

This type of meeting could be accomplished through either augmented reality (where you superimpose a digital layer on top of our physical environment) or virtual reality (where you enter a completely immersive world). The change won't come right away, since most people don't own tools to enable this kind of capture yet, in contrast to the way the switch to video meetings was enabled by the fact that many people already had PCs or phones with cameras. Right now, you can use virtual reality goggles and gloves to control your avatar, but more sophisticated and less obtrusive tools—like lightweight glasses and contact lenses—will come along over the next few years.

Improvements in computer vision, display technology, audio, and sensors will capture your facial expressions, eyeline, and body language with very little delay. Think about any time you've tried to jump in with a thought during a spirited video meeting, and how hard that was to do when you couldn't see the way people's body language shifts as they're wrapping up a thought.

A key feature in the metaverse is the use of spatial audio, which makes speech sound like it's actually coming from the direction of the person talking. True presence means that technology captures what it *feels* like to be in a room with someone, not just what it *looks* like.

In the fall of 2021, I got to put on a headset and join a meeting in the metaverse. It was amazing to hear how people's voices seemed to move along with them. You don't realize how unusual it is to have meeting audio only coming from your computer's speaker until you try something else. In the metaverse, you'll be able to lean over and have a quiet side conversation with a coworker just as if you're in the same room.

I'm particularly excited to see how metaverse technologies will enable more spontaneity with remote work. This is the biggest thing you lose when you're not in the office. Working from your living room isn't exactly conducive to having an unplanned discussion with your manager about your last meeting or starting a casual conversation with your new coworker about last night's baseball game. But if you're all working together remotely in a virtual space, you'll be able to see when someone is free and approach that person to chat.

We're nearing a threshold at which the technology is beginning to truly replicate the experience of being in the office. The changes we've seen in the workplace are precursors to changes that I think we'll eventually see in many areas. We're moving toward a future where we will all spend more time around and within digital spaces. The metaverse may feel like a novel concept now, but as technology gets better, it will evolve into what feels more like an extension of our physical world.

There are, of course, huge sectors of the economy where workplaces won't change as much or will shift in different ways from what I'm describing here. If you're a flight attendant, your job has probably evolved a lot in recent years but not because of increased digitization. If you're a server in a restaurant, your customers might now use a QR code menu to decide what they want before placing orders through their phones. And if you work on a factory floor, technology has been changing your job since long before the pandemic.*

Digitization will eventually transform all of our lives in one way or another, though. Consider how the way you take care of your health may have changed since 2020. Have any of your doctor's appointments in the past couple of years been virtual? Had you ever done a virtual health appointment before COVID? The number of people using telehealth services increased by a factor of 38 during the pandemic.

The benefits of telehealth are clear during a disease outbreak. People who might have been skeptical about virtual appointments before suddenly saw a tangible upside to them: If you're not feeling well, it's a lot safer to do your appointment from home where you don't have to worry about infecting anyone or getting infected yourself.

Once you try telehealth, though, it becomes clear that the ben-

* In addition to the rise of automation, augmented reality is catching on so that workers can be trained in complex tasks and quickly see the status of a piece of equipment with just a glance.

efits go way beyond limiting your exposure to people who are sick. Going to the doctor can be a time-consuming activity, as you have to take time off work or find someone to watch your kids, travel to the doctor's office, sit in a waiting room, check out after your appointment, and then commute back home or to work. That might be worth it for some types of visits, but it feels increasingly unnecessary for others—especially behavioral health visits.

Seeing your therapist is a lot less time-consuming and easier to fit into your day when you only need to turn a laptop on. Sessions can be as long or as short as needed. A fifteen-minute session might not feel worth it if you have to go to someone's office, but it makes a lot more sense from home. Plus, many people feel more comfortable in their own space than in a clinical setting.

Other types of doctor's visits might also become more flexible as new tools emerge. Right now, when it's time for your annual physical, you probably need to go into your doctor's office to get your vitals taken and your blood drawn—but what if you had a private, secure device at home that your doctor could control remotely to test your blood pressure?

One day soon, your doctor might be able to look at data collected from a smartwatch—with your permission—to see how you're sleeping and how your active heart rate differs from your resting heart rate. Instead of going into an office to get blood drawn, you could get your blood tested instead at a convenient place in your neighborhood—maybe at your local pharmacy—that sends the results directly to your doctor. And if you move to another state, you could still see the same primary care physician you've trusted for years.

These are all real possibilities in the future. There will always be health care specialties that require in-person visits—I can't imagine a future where a robot removes your appendix in your living room—but most routine care will eventually be something you can do from the comfort of your own home.

—

I don't think that virtual alternatives will replace existing structures in K–12 education the way they are likely to in office work and health care. But change is coming to education all the same. Although the COVID pandemic made it clear that young people learn best when they can work face-to-face with their teachers, digitization will lead to new tools that supplement what happens in the classroom.

If you were the parent of a school-age child during the pandemic, chances are you became very familiar with the concepts of synchronous and asynchronous learning. Synchronous learning attempts to mimic the normal experience of going to school: A teacher uses a videoconferencing service to teach a class live, and students can jump in to ask questions, just like in a real classroom. This will remain a good option for many postsecondary students, especially those who require more flexibility. But I don't see synchronous K–12 learning sticking around much in a post-pandemic world except maybe for the oldest high school students or on snow days. It just doesn't work well for younger students.

Asynchronous learning, on the other hand, is here to stay—although in a different form from what we saw during the height of the pandemic. In this type of instruction students watch prerecorded lectures and complete assignments on their own schedule, and teachers can post prompts on a discussion board and ask their class to weigh in for credit.

I know that both forms of remote learning were frustrating for many teachers, parents, and students, and that the idea of keeping any version of them might not seem appealing. But there's remarkable potential for some of the tools used in asynchronous learning to supplement the work that students and teachers already do together in the classroom.

Just think about how digital curricula can make homework more enriching and engaging. If you're a student, you'll be able to get feedback in real time while you do your homework online. The days of turning in your homework and then waiting to see what you got right should be behind you. The content will be more interactive and personalized to you, helping you focus on areas where you need a bit more help while boosting your confidence by giving you problems you're comfortable solving.

If you're a teacher, you'll be able to see how quickly your students worked and how often they needed hints, giving you a deeper understanding of how they are doing. A simple button click might show you that Noah needs more help on a particular type of question or that Olivia is ready to take on a more advanced reading assignment.

Digital tools can facilitate more personalized learning in the classroom too. One example I'm familiar with is the Summit Learning Platform. Students work together with teachers to pick a goal—maybe you want to get into a specific university or prepare for a certain career path—and create a digital learning plan. In addition to receiving traditional instruction in the classroom, they use the platform to test their knowledge and assess their own performance. Letting kids take control of their learning in this way helps build confidence, curiosity, and persistence.

These technologies have been in the works for a while, but progress was accelerated when demand skyrocketed during the pandemic. In the years ahead, the Gates Foundation will invest heavily in these tools and measure what works.

Some of the biggest leaps forward have been in math curricula—especially algebra. Algebra I is a key milestone on the road to graduation, but it has the highest failure rate of any high school course. Students who don't pass have just a one-in-five chance of graduating from high school, a problem that particularly affects students who are Black, Latino, English language learners, or experiencing poverty, putting them at a disadvantage for future careers and higher earnings. Kids who struggle with algebra often develop a self-image of not being good at math that haunts them through the rest of their time in school. They get frustrated by problem sets that are perhaps too hard for their current skill level and never catch up as classes become more advanced.

One example of a company that's working on digitally enabled innovation is Zearn. Its new math curriculum for grade school students helps them shore up concepts that are key to more advanced math, like fractions and the order of operations. They provide teaching materials to help educators create lesson plans, and they've created digital lessons and assignments that make doing homework more fun.

I'm hopeful that tools like this will help more students succeed in school while also easing the burden on teachers. Unlike the way things were at the height of the pandemic—when remote learning meant that teachers had to juggle more work than normal—software will eventually free up more time for them to focus on where they add the most value.

Of course, the ability of new digital education tools to transform learning is dependent on kids having access to technology at home. The gap has narrowed since the start of the pandemic and will continue to narrow, but a lot of kids still don't have a decent computer or reliable, fast internet access. (This is especially true for students of color and those from low-income families, who stand to benefit the most from digital tools that can help close educational outcome gaps.) Finding ways to expand access is just as important as the development of new innovations. Ultimately, the extent to which digitization takes hold—whether it's in education or any other area—depends on how widespread adoption is.

In 1964, Bell Telephone exhibited the first-ever video phone at the World's Fair. The Picturephone looked like something from *The Jetsons,* with a small live image embedded in a futuristic-looking oval tube. I was eight years old at the time. I saw pictures of the phone

Virtual meetings have come a long way since this early prototype of Bell Telephone's Picturephone in 1964.

in the newspaper and couldn't believe that what I was looking at was possible. Little did I know that, decades later, I'd spend hours of my day on video calls.

It's easy to see technology as mundane when it's just part of our day-to-day life. When you take the time to think about them, though, today's digital capabilities are miraculous. We're now able to connect with one another and with the world in a way that once seemed like pure fantasy.

For many people—especially older people in assisted-living facilities—video calls have become a lifeline to the world. Even if you're tired of virtual happy hours and birthday parties, you can't deny that the connections they've provided helped get us through the darkest days of the pandemic.

As devastating as the COVID pandemic was, imagine how much worse the isolation would have been even a decade ago. Video calls existed, but broadband speeds weren't fast enough yet to support lots of people doing video meetings from home. The reason broadband infrastructure improved so rapidly over the past decade was that people wanted to be able to watch Netflix at night. By the time the pandemic started, bandwidth had increased enough to let people work remotely during the daytime.

The truth is, it's impossible to predict exactly how breakthroughs will shape the future. You can come up with all these scenarios for ways in which a new technology will change the world, and then something like COVID comes along and forces everyone to use the tools at their disposal in new ways. For all her amazing foresight, I doubt even Katalin Karikó imagined that mRNA vaccines would one day play an essential role in ending a pandemic.

I can't wait to see how digital breakthroughs continue to evolve in the years ahead. The technological advances we've seen over the past couple of years have the potential to create more flexibility and options that improve people's lives. They'll even put us in a better position to prevent the next pandemic. When we look back at this period, I suspect history will view it as a time of terrible devastation and loss that also sparked massive changes for the better.

GLOSSARY

Antibodies: Proteins created by the immune system that grab onto the surface of a pathogen and attempt to neutralize it.

Antigen test: A disease diagnostic that looks for specific proteins on the surface of a pathogen. Antigen tests are marginally less accurate than PCR tests but provide fast results, don't require a lab, and are good at identifying when an infected person may be contagious. Lateral flow immunoassays—the kind that resemble home pregnancy tests—are antigen tests.

Breakthrough infection: An infection that occurs in a person who has been vaccinated against a disease.

CEPI: The Coalition for Epidemic Preparedness Innovations, a nonprofit created in 2017 to accelerate work on vaccines against new infectious diseases and help those vaccines reach people in the poorest countries.

Cold chain: The process of keeping a vaccine at the right temperature as it travels from the factory where it is made to the place where it is being administered.

Contact tracing: The process of identifying people who came into contact with someone who was infected with a certain disease.

COVAX: The global effort to get COVID vaccines out to low- and middle-income countries, co-led by CEPI, Gavi, and the WHO.

Effectiveness, efficacy: The measure of how well a vaccine or drug works. In the medical field, *efficacy* refers to performance in a clinical trial, and effectiveness refers to performance in the real world. For the sake of simplicity, in this book I've used *effectiveness* to refer to both.

Gavi, the Vaccine Alliance: A nonprofit created in 2000 to encourage manufacturers to lower vaccine prices for the poorest countries in return for long-term, high-volume, and predictable demand from those countries. Formerly known as the Global Alliance for Vaccines and Immunization.

Genome, genomic sequencing: The genome is the genetic code of an organism. All living things have genomes, and every genome is unique. Sequencing a pathogen's genome is the process of figuring out the order in which its genetic information appears.

GERM: The Global Epidemic Response and Mobilization team. A proposed global organization responsible for detecting and responding to outbreaks and preventing them from becoming pandemics.

Global Fund: Officially the Global Fund to Fight AIDS, TB, and Malaria, a nonprofit partnership designed to end the epidemics of those three diseases.

IHME: The Institute for Health Metrics and Evaluation, a research organization based at the University of Washington that develops evidence to guide decisions about public health.

Monoclonal antibodies (mAbs): A form of treatment for some diseases. These are antibodies that have been isolated from a patient's blood or designed in a lab and then cloned billions of times to create a treatment for someone who has been infected.

mRNA (messenger RNA): Genetic material that carries the directions for making certain proteins to the factories in your cells where the proteins will be assembled. Vaccines that use mRNA work by introducing genetic code that teaches your cells to make shapes that match certain shapes on a given virus, triggering your immune system to produce antibodies against that virus.

Nonpharmaceutical interventions (NPIs): Policies and tools that reduce the spread of an infectious disease without the use of vaccines or drugs. Common NPIs include masks, social distancing, quarantines, business and school closures, travel restrictions, and contact tracing.

PCR test: Polymerase chain reaction, the current gold standard in disease diagnostics.

SCAN: The Seattle Coronavirus Assessment Network, which along with the Seattle Flu Study was set up to learn more about how a respiratory disease spreads through a community.

WHO: The World Health Organization, a division of the United Nations responsible for international public health.

ACKNOWLEDGMENTS

I want to thank all of the staff, trustees, grantees, and partners of the Bill & Melinda Gates Foundation who worked tirelessly to help out during COVID. I'm inspired by your passion and commitment. Melinda and I are lucky to work with such a talented group of people.

Writing this book was like trying to hit a moving target, as new information came in almost daily. So it took a team effort to stay on top of the latest data and analysis. I'm grateful to everyone who helped me complete *How to Prevent the Next Pandemic.*

I have written each of my books with one or more writing and research partners. For this book, as he did on my previous one, Josh Daniel devoted his considerable skill to helping me explain complicated topics simply and clearly. Josh and his colleagues Paul Nevin and Casey Selwyn made a fantastic trio who conducted in-depth research, synthesized ideas from experts across many fields, and helped me clarify my thinking. I appreciate their advice and admire their hard work.

For this book, I benefited from the insights of many people at the foundation, including Mark Suzman, Trevor Mundel, Chris Elias, Gargee Ghosh, Anita Zaidi, Scott Dowell, Dan Wattendorf, Lynda Stuart, Orin Levine, David Blazes, Keith Klugman, and Susan Byrnes. They joined brainstorming sessions and reviewed drafts while balancing the rest of their demanding jobs during a pandemic. Many others at the foundation provided expert input, research, and feedback on drafts, including Hari Menon, Oumar Seydi, Zhi-Jie Zheng, Natalie Africa, Mary Aikenhead, Jennifer Alcorn, Valerie Nkamgang Bemo, Adrien de Chaisemartin, Jeff Chertack, Chris Culver, Emily Dansereau, Peter Dull, Ken Duncan, Emilio Emini, Mike Famulare, Michael Galway, Allan Golston, Vishal Gujadhur, Dan Hartman, Vivian Hsu, Hao Hu, Emily Inslee, Carl Kirkwood, Dennis Lee, Murray Lumpkin, Barbara Mahon, Helen Matzger, Georgina Murphy, Rob Nabors, Natalie Revelle, David Robinson, Torey de Rozario, Tanya Shewchuk, Duncan Steele, Katherine Tan, Brad Tytel, David Vaughn, Philip Welkhoff, Edward Wenger, Jay Wenger, Greg Widmyer, and Brad Wilken. And the foundation's communications and advocacy teams not only contributed research but will carry this work forward, helping me translate the ideas in this book into concrete changes that leave the world more prepared to deal with the next major outbreak.

Thoughtful reviews of early passages and drafts came from Anthony Fauci, David Morens, Tom Frieden, Bill Foege, Seth Berkley, Larry Brilliant, Sheila Gulati, and Brad Smith.

I also want to thank the many people at Gates Ventures who helped make this book possible.

Larry Cohen provided leadership and vision that are both essential and rare. I appreciate his calm demeanor, wise guidance, and dedication to the work we do together.

Niranjan Bose gave me expert advice and helped me get many technical details right. Becky Bartlein and the rest of the Exemplars in Global Health team helped me flesh out details on why some countries did so much better than others.

Alex Reid thoughtfully drove the communications team that was responsible for ensuring the successful launch of the book. Joanna Fuller was instrumental in helping me with all the details of the story of the Seattle Flu Study and SCAN.

Andy Cook led the online strategy work that brought the book out online on my website, social channels, and beyond.

Ian Saunders did a masterful job leading the creative team that helped bring the book to market.

Meghan Groob offered sound editorial advice, particularly on the Afterword. Anu Horsman led the creative process for the visual content of the book. Jen Krajicek worked behind the scenes to manage its production. Brent Christofferson oversaw the production of the visual assets with charts from Beyond Words and illustrations from Jono Hey. John Murphy helped me identify and learn about many of the heroes of the COVID fight.

Greg Martinez and Jennie Lyman help me stay up-to-date on where technology is headed, work that informed the Afterword in particular.

Gregg Eskenazi and Laura Ayers negotiated contracts and secured permissions from dozens of sources featured in this book.

Many others played an important role in the creation and release of this book, including Katie Rupp, Kerry McNellis, Mara MacLean, Naomi Zukor, Cailin Wyatt, Chloe Johnson, Tyler Hughes, Margaret Holsinger, Josh Friedman, Ada Arinze, Darya Fenton, Emily Warden, Zephira Davis, Khiota Therrien, Abbey Loos, K.J. Sherman, Lisa Bishop, Tony Hoelscher, Bob Regan, Chelsea Katzenberg, Jayson Wilkinson, Maheen Sahoo, Kim McGee, Sebastian Majewski, Pia Dierking, Hermes Arriola, Anna Dahlquist, Sean Williams, Bradley Castaneda, Jacqueline Smith, Camille Balsamo-Gillis, and David Sanger.

And I want to thank the rest of the incredible team at Gates Ventures: Aubree Bogdonovich, Hillary Bounds, Patrick Brannelly, Gretchen Burk, Maren Claassen, Matt Clement, Quinn Cornelius, Alexandra Crosby, Prarthna Desai, Jen Kidwell Drake, Sarah Fosmo, Lindsey Funari, Nathaniel Gerth, Jonah Goldman, Andrea Vargas Guerra, Rodi Guidero, Rob Guth, Rowan Hussein, Jeffrey Huston, Gloria Ikilezi, Farhad Imam, Tricia Jester, Lauren Jiloty,

Goutham Kandru, Sarah Kester, Liesel Kiel, Meredith Kimball, Jen Langston, Siobhan Lazenby, Anne Liu, Mike Maguire, Kristina Malzbender, Amelia Mayberry, Caitlin McHugh, Emma McHugh, Angelina Meadows, Joe Michaels, Craig Miller, Ray Minchew, Valerie Morones, Henry Moyers, Dillon Mydland, Kyle Nettelbladt, Bridgette O'Connor, Patrick Owens, Dreanna Perkins, Mukta Phatak, David Vogt Phillips, Tony Pound, Shirley Prasad, Zahra Radjavi, Kate Reizner, Chelsea Roberts, Brian Sanders, Bennett Sherry, Kevin Smallwood, Steve Springmeyer, Aishwarya Sukumar, Jordan-Tate Thomas, Alicia Thompson, Caroline Tilden, Rikki Vincent, Courtney Voigt, William Wang, Stephanie Williams, Sunrise Swanson Williams, Tyler Wilson, Sydney Yang, Jamal Yearwood, and Mariah Young.

A special thank-you to the human resources teams at both Gates Ventures and the Gates Foundation for all they've done during COVID to maintain a strong culture while putting everyone's health and safety first.

Chris Murray and the rest of the team at the Institute for Health Metrics and Evaluation assisted with research, modeling, and analysis that informed my thinking as well as many of the charts and statistics in this book.

Max Roser's site Our World in Data is an invaluable resource, and I turned to it countless times while writing this book.

This book would not have been possible without the tireless support of my editor, Robert Gottlieb at Knopf. His guidance helped us keep the book clear and reader-friendly. Katherine Hourigan masterfully managed the entire process, helping us stay on track under a tight (self-imposed) deadline. And I want to thank everyone else at Penguin Random House who supported this book: Reagan Arthur, Maya Mavjee, Anne Achenbaum, Andy Hughes, Ellen Feldman, Mike Collica, Chris Gillespie, Erinn Hartman, Jessica Purcell, Julianne Clancy, Amy Hagedorn, Laura Keefe, Suzanne Smith, Serena Lehman, and Kate Hughes.

Warren Buffett's incredibly generous support for the Gates Foundation, a pledge he first made in 2006, has allowed us to expand and deepen our work around the world. I'm honored by his commitment and feel fortunate to call him my friend.

I've learned a great deal from Melinda since that day we met in 1987. I'm deeply proud of the family we raised together and the foundation we created together.

Finally, I want to thank Jenn, Rory, and Phoebe. The year in which I wrote this book was an incredibly difficult one for the world and, personally, for our family. I am grateful for their constant support and love. Nothing means more to me than being their dad.

NOTES

Introduction

3 The Chinese government had taken: Hien Lau et al., "The Positive Impact of Lockdown in Wuhan on Containing the COVID-19 Outbreak in China," *Journal of Travel Medicine* 27, no. 3 (April 2020).

5 Nick reported that diarrhea was killing: Nicholas D. Kristof, "For Third World, Water Is Still a Deadly Drink," *New York Times,* Jan. 9, 1997.

5 Photo: From *The New York Times.* © 1997 The New York Times Company. All rights reserved. Used under license.

6 One of the most influential: World Bank, World Development Report 1993, https://elibrary.worldbank.org.

6 There were 1.5 million new cases: World Health Organization (WHO), "Number of New HIV Infections," https://www.who.int.

7 It struck Madagascar in 2017: "Managing Epidemics: Key Facts About Major Deadly Diseases," WHO, 2018, https://who.int.

7 Figure: Endemic killers. Source: Institute for Health Metrics and Evaluation (IHME) at the University of Washington, Global Burden of Disease Study 2019, https://healthdata.org.

7 In 2000, these diseases killed: Institute for Health Metrics and Evaluation, GBD Compare, https://vizhub.healthdata.org/gbd-compare/.

8 Photo: Eye Ubiquitous/Universal Images Group via Getty Images.

9 Photo: Fototeca Storica Nazionale via Getty Images.

9 In 2019, before COVID, tourists: Our World in Data, "Tourism," https://www.ourworldindata.org.

11 By that July: "2014–2016 Ebola Outbreak in West Africa," Centers for Disease Control and Prevention (CDC), https://www.cdc.gov.

12 Photo: Enrico Dagnino/*Paris Match* via Getty Images.

20 In 2021, the White House announced: Seth Borenstein, "Science Chief Wants Next Pandemic Vaccine Ready in 100 Days," Associated Press, June 2, 2021.

21 Every year, influenza alone: WHO, "Global Influenza Strategy 2019–2030," https://www.who.int.

Chapter 1. Learn from COVID

24 Figure: The true toll of COVID. Estimated number of global excess deaths include official count of COVID-19 deaths, additional estimated COVID-19 deaths, and deaths from all causes attributed to complications stemming from the pandemic through December 2021. Source: Institute for Health Metrics and Evaluation (IHME) at the University of Washington (2021).

25 Toward the end of 2021: Our World in Data, "Estimated Cumulative Excess Deaths Per 100,000 People During COVID-19," https://ourworldindata.org/.

25 Figure: Containing COVID in Vietnam. New cases per day (seven-day rolling average). Source: "Emerging COVID-19 Success Story: Vietnam's Commitment to Containment," Exemplars in Global Health program, https://www.exemplars.health (published March 2021; accessed Jan. 2022). Using data extracted from Hannah Ritchie et al., "Coronavirus Pandemic (COVID-19)" (2020), published online at OurWorldInData.org, https://ourworldindata.org/coronavirus.

25 Its rate of excess deaths: Our World in Data, "Estimated Cumulative Excess Deaths per 100,000 People During COVID-19," https://ourworldindata.org.

26 IHME's data also suggests: T. J. Bollyky et al., "Pandemic Preparedness and COVID-19: An Exploratory Analysis of Infection and Fatality Rates, and Contextual Factors Associated with Preparedness in 177 Countries, from January 1, 2020, to September 30, 2021," *The Lancet,* in press.

27 Uganda and its neighbors: Prosper Behumbiize, "Electronic COVID-19 Point of Entry Screening and Travel Pass DHIS2 Implementation at Ugandan Borders," https://community.dhis2.org.

27 Photo: Sally Hayden/SOPA Images/LightRocket via Getty Images.

30 Photo: The Gates Notes, LLC/Ryan Lobo.

31 To protect her family: "7 Unsung Heroes of the Pandemic," *Gates Notes,* https://gatesnotes.com.

31 Around the world, health care workers: WHO, "Health and Care Worker Deaths During COVID-19," https://www.who.int.

36 And expecting perfection: This account of David Sencer's experience is based on this interview: Victoria Harden (interviewer) and David Sencer (interviewee), CDC, "SENCER, DAVID J.," *The Global Health Chronicles*, https://globalhealthchronicles.org/ (accessed Dec. 28, 2021).

38 In total, GBS cases occurred: Kenrad E. Nelson, "Invited Commentary: Influenza Vaccine and Guillain-Barré Syndrome—Is There a Risk?," *American Journal of Epidemiology* 175, no. 11 (June 1, 2012): 1129–32.

39 In 2021 alone, they accounted for: UNICEF, "COVID-19 Vaccine Market Dashboard," https://www.unicef.org; and data provided by Linksbridge.

40 But as the late educator and physician: Hans Rosling, *Ten Reasons We're Wrong About the World—and Why Things Are Better Than You Think* (Flatiron Books, 2018).

Chapter 2. Create a pandemic prevention team

42 In the year 6 CE: Michael Ng, "Cohorts of Vigiles," in *The Encyclopedia of the Roman Army* (2015): 122–276.

42 In America, there were volunteer groups: Merrimack Fire, Rescue, and EMS, "The History of Firefighting," https://www.merrimacknh.gov/about-fire-rescue.

43 There are now about: U.S. Bureau of Labor Statistics, "Occupational Employment and Wages, May 2020," https://www.bls.gov/; National Fire Protection Association, "U.S. Fire Department Profile 2018," https://www.nfpa.org.

43 For nearly 800 years: Thatching Info, "Thatching in the City of London," https://www.thatchinginfo.com/.

43 Today, one large fire-prevention nonprofit: National Fire Protection Association, https://www.nfpa.org.

48 An Egyptian tablet: Global Polio Eradication Initiative (GPEI), "History of Polio," https://www.polioeradication.org/.

49 By adding a vaccine for it: GPEI, https://www.polioeradication.org.

49 Figure: Ending polio. Data plotted are wild polio cases only. Sources: WHO, Progress Towards Global Immunization Goals, 2011 (accessed Jan. 2022), data provided by 194 WHO Member States.

50 Photo: © UNICEF/UN0581966/Herwig.

51 The coordinator of Pakistan's: Interview with Dr. Shahzad Baig, National Coordinator, Pakistan National Emergency Operation Centre, July 2021.

51 To put that number in perspective: IISS, "Global Defence-Spending on the Up, Despite Economic Crunch," https://www.iiss.org.

Chapter 3. Get better at detecting outbreaks early

55 Countries in Africa, for example: CDC, "Integrated Disease Surveillance and Response (IDSR)," https://www.cdc.gov.

57 In Vietnam, teachers are trained: A. Clara et al., "Developing Monitoring and Evaluation Tools for Event-Based Surveillance: Experience from Vietnam," *Global Health* 16, no. 38 (2020).

57 According to the WHO: "Global Report on Health Data Systems and Capacity, 2020," https://www.who.int.

58 At the end of October 2021: IHME, "Global COVID-19 Results Briefing," Nov. 3, 2021, https://www.healthdata.org.

58 In Europe the rate was: IHME results briefings for the European Union and Africa, https://healthdata.org.

58 Over time, the Gates Foundation: Estimates generated by the Vaccine Impact Modeling Consortium based on its publication by Jaspreet Toor et al., "Lives Saved with Vaccination for 10 Pathogens Across 112 Countries in a Pre-COVID-19 world," July 13, 2021.

59 Mozambique is also one of several: CHAMPS, "A Global Network Saving Lives," https://champshealth.org.

59 In 2013, we funded researchers: MITS Alliance, "What Is MITS?," https://mitsalliance.org.

60 Photo: The Gates Notes, LLC/Curator Pictures, LLC.

62 Because you already know: Cormac Sheridan, "Coronavirus and the Race to Distribute Reliable Diagnostics," *Nature Biotechnology* 38 (April 2020): 379–91.

65 Amazingly, the Nexar system can process: LGC, Biosearch Technologies, Nexar technical specs, https://www.biosearchtech.com.

65 Photo: LGC, Biosearch Technologies™.

69 Most ingredients in an ordinary salad: Email correspondence with Lea Starita of the Advanced Technology Lab at Brotman Baty Institute.

73 Figure: When COVID arrived in Washington state. Data accessed Dec. 9, 2021. Confirmed daily infections represent reported cases per day. Estimated infections are the number of people estimated to be infected with COVID-19 each day, including those not tested. COVID data available between Feb. 2020 and April 1, 2020. Source: Institute for Health Metrics and Evaluation (IHME) at the University of Washington.

74 "Coronavirus May Have Spread": Sheri Fink and Mike Baker, "Coronavirus May Have Spread in U.S. for Weeks, Gene Sequencing Suggests," *New York Times,* March 1, 2020.

77 Oxford Nanopore is now working with: Oxford Nanopore, "Oxford Nanopore, the Bill and Melinda Gates Foundation, Africa Centres for Disease

Control and Prevention and Other Partners Collaborate to Transform Disease Surveillance in Africa," https://nanoporetech.com.

80 In March 2020, Neil Ferguson: Neil M. Ferguson et al., "Report 9—Impact of Non-Pharmaceutical Interventions (NPIs) to Reduce COVID-19 Mortality and Healthcare Demand," https://www.imperial.ac.uk.

Chapter 4. Help people protect themselves right away

84 In her book *On Immunity:* Bill Gates, "Where Do Vaccine Fears Come From?," https://www.gatesnotes.com.

85 Photo: Gado via Getty Images.

88 One study found: Steffen Juranek and Floris T. Zoutman, "The Effect of Non-Pharmaceutical Interventions on the Demand for Health Care and on Mortality: Evidence from COVID-19 in Scandinavia," *Journal of Population Economics* (July 2021): 1–22, doi:10.1007/s00148-021-00868-9.

88 Another study estimated: Solomon Hsiang et al., "The Effect of Large-Scale Anti-Contagion Policies on the COVID-19 Pandemic," *Nature* 584, no. 7820 (Aug. 2020): 262–67, doi:10.1038/s41586-020-2404-8.

90 Between March 2020 and June 2021: UNESCO, "School Closures and Regional Policies to Mitigate Learning Losses in Asia Pacific," https://uis.unesco.org.

91 Figure: COVID is much worse for older people. Estimated infection fatality ratio (%) includes estimated number of people who died from COVID-19 globally in 2020, both sexes, prior to vaccination introduction. Source: Institute for Health Metrics and Evaluation (IHME) at the University of Washington.

91 The United Nations estimates: UNESCO.

91 In the United States: Emma Dorn et al., "COVID-19 and Learning Loss—Disparities Grow and Students Need Help," McKinsey & Company, Dec. 8, 2020, https://www.mckinsey.com.

92 In the United States through March 2021: CDC, "Science Brief: Transmission of SARS-CoV-2 in K–12 Schools and Early Care and Education Programs—Updated," Dec. 2021, https://www.cdc.gov.

93 One study found that: Victor Chernozhukov et al., "The Association of Opening K–12 Schools with the Spread of COVID-19 in the United States: County-Level Panel Data Analysis," *Proceedings of the National Academy of Sciences* (Oct. 2021): 118.

94 One ingenious study used: Joakim A. Weill et al., "Social Distancing Responses to COVID-19 Emergency Declarations Strongly Differentiated by Income," *Proceedings of the National Academy of Sciences of the United States of America* (Aug. 2020): 19658–60.

95 Every year, influenza kills: CDC, "Frequently Asked Questions About Estimated Flu Burden," https://www.cdc.gov; WHO, "Ask the Expert: Influenza Q&A," https://www.who.int.

97 According to the journal *Nature:* "Why Many Countries Failed at COVID Contact-Tracing—but Some Got It Right," *Nature,* Dec. 14, 2020.

97 In March 2020: Ha-Linh Quach et al., "Successful Containment of a Flight-Imported COVID-19 Outbreak Through Extensive Contact Tracing, Systematic Testing and Mandatory Quarantine: Lessons from Vietnam," *Travel Medicine and Infectious Disease* 42 (Aug. 2021).

98 In two counties: R. Ryan Lash et al., "COVID-19 Contact Tracing in Two Counties—North Carolina, June–July 2020," *MMWR: Morbidity and Mortality Weekly Report* 69 (Sept. 25, 2020).

98 This daily testing: B. C. Young et al., "Daily Testing for Contacts of Individuals with SARS-CoV-2 Infection and Attendance and SARS-CoV-2 Transmission in English Secondary Schools and Colleges: An Open-Label, Cluster-Randomised Trial," *The Lancet* (Sept. 2021).

98 In general, if you start: Billy J. Gardner and A. Marm Kilpatrick, "Contact Tracing Efficiency, Transmission Heterogeneity, and Accelerating COVID-19 Epidemics," *PLOS Computational Biology* (June 17, 2021).

99 About 70 percent of those: Dillon C. Adam et al., "Clustering and Superspreading Potential of SARS-CoV-2 Infections in Hong Kong," *Nature Medicine* (Sept. 2020).

99 For reasons we don't entirely understand: Kim Sneppen et al., "Overdispersion in COVID-19 Increases the Effectiveness of Limiting Nonrepetitive Contacts for Transmission Control," *Proceedings of the National Academy of Sciences of the United States of America* 118, no. 14 (April 2021).

99 Understanding this limitation: W. J. Bradshaw et al., "Bidirectional Contact Tracing Could Dramatically Improve COVID-19 Control," *Nature Communications* (Jan. 2021).

100 One study found: Akira Endo et al., "Implication of Backward Contact Tracing in the Presence of Overdispersed Transmission in COVID-19 Outbreaks," *Wellcome Open Research* 5, no. 239 (2021).

101 In Sydney, Australia: Anthea L. Katelaris et al., "Epidemiologic Evidence for Airborne Transmission of SARS-CoV-2 During Church Singing, Australia, 2020," *Emerging Infectious Diseases* 27, no. 6 (2021): 1677.

101 At a restaurant in Guangzhou, China: Jianyun Lu et al., "COVID-19 Outbreak Associated with Air Conditioning in Restaurant, Guangzhou, China, 2020," *Emerging Infectious Diseases* 26, no. 7 (2020): 1628.

101 In Christchurch, New Zealand: Nick Eichler et al., "Transmission of Severe Acute Respiratory Syndrome Coronavirus 2 During Border Quarantine and Air Travel, New Zealand (Aotearoa)," *Emerging Infectious Diseases* 27, no. 5 (2021): 1274.

102 In fact, even if someone: CDC, "Science Brief: SARS-CoV-2 and Surface (Fomite) Transmission for Indoor Community Environments," April 2021, https://www.cdc.gov.

102 And for a time, at least: Apoorva Mandavilli, "Is the Coronavirus Getting Better at Airborne Transmission?," *New York Times,* Oct. 1, 2021.

103 One study used a computer simulation: Rommie Amaro et al., "#COVID isAirborne: AI-Enabled Multiscale Computational Microscopy of Delta SARS-CoV-2 in a Respiratory Aerosol," Nov. 17, 2021, https://sc21.super computing.org.

105 It dates from 1910: Christos Lynteris, "Why Do People Really Wear Face Masks During an Epidemic?," *New York Times,* Feb. 13, 2020; Wudan Yan, "What Can and Can't Be Learned from a Doctor in China Who Pioneered Masks," *New York Times,* May 24, 2021.

106 For many people: M. Joshua Hendrix et al., "Absence of Apparent Transmission of SARS-CoV-2 from Two Stylists After Exposure at a Hair Salon with a Universal Face Covering Policy—Springfield, Missouri, May 2020," *Morbidity and Mortality Weekly Report* 69 (2020): 930–32.

106 The first is called: J. T. Brooks et al., "Maximizing Fit for Cloth and Medical Procedure Masks to Improve Performance and Reduce SARS-CoV-2 Transmission and Exposure," *Morbidity and Mortality Weekly Report* 70 (2021): 254–57.

107 One research team: Siddhartha Verma et al., "Visualizing the Effectiveness of Face Masks in Obstructing Respiratory Jets," *Physics of Fluids* 32, no. 061708 (2020).

107 Another group of researchers: J. T. Brooks et al., "Maximizing Fit for Cloth and Medical Procedure Masks to Improve Performance and Reduce SARS-CoV-2 Transmission and Exposure," *Morbidity and Mortality Weekly Report* 70 (2021): 254–57.

107 One study found: Gholamhossein Bagheri et al., "An Upper Bound on One-to-One Exposure to Infectious Human Respiratory Particles," *Proceedings of the National Academy of Sciences* 118, no. 49 (Dec. 2021).

108 Photo: The Gates Notes, LLC/Sean Williams.

109 In San Francisco: Christine Hauser, "The Mask Slackers of 1918," *New York Times,* Dec. 10, 2020.

109 And in Bangladesh, researchers: Jason Abaluck et al., "Impact of Community Masking on COVID-19: A Cluster-Randomized Trial in Bangladesh," *Science,* Dec. 2, 2021.

Chapter 5. Find new treatments fast

111 Its director-general said: Tedros Adhanom Ghebreyesus, remarks at the Munich Security Conference, Feb. 15, 2020, https://www.who.int.

111 In just the first half of 2020 alone: WHO, "Coronavirus Disease (COVID-19) Advice for the Public: Mythbusters," May 2021, https://www.who.int; Ian Freckelton, "COVID-19: Fear, Quackery, False Representations and the Law," *International Journal of Law and Psychiatry* 72, no. 101611 (Sept.–Oct. 2020).

113 Hundreds of clinical studies: U.S. National Library of Medicine, https://clinicaltrials.gov (search for "COVID-19 and hydroxychloroquine"); Peter Horby and Martin Landray, "No Clinical Benefit from Use of Hydroxychloroquine in Hospitalised Patients with COVID-19," June 5, 2020, https://www.recoverytrial.net.

113 Meanwhile, the hydroxychloroquine craze: Aliza Nadi, " 'Lifesaving' Lupus Drug in Short Supply After Trump Touts Possible Coronavirus Treatment," NBC News, March 23, 2020.

113 By that summer: The Recovery Collaborative Group, "Dexamethasone in Hospitalized Patients with Covid-19," *New England Journal of Medicine,* Feb. 25, 2021.

114 Less than a month after: Africa Medical Supplies Platform, July 17, 2020, https://amsp.africa; Ruth Okwumbu-Imafidon, "UNICEF in Negotiations to Buy COVID-19 Drug for 4.5 Million Patients in Poor Countries," *Nairametrics,* July 30, 2020.

114 British researchers estimated: England National Health Service, "COVID Treatment Developed in the NHS Saves a Million Lives," March 23, 2021, https://www.england.nhs.uk.

115 However, a subsequent study: Robert L. Gottlieb et al., "Early Remdesivir to Prevent Progression to Severe Covid-19 in Outpatients," *New England Journal of Medicine,* Dec. 22, 2021.

116 If you're infected: U.S. National Institutes of Health, "Table 3a. Anti-SARS-CoV-2 Monoclonal Antibodies: Selected Clinical Data," Dec. 2021, https://www.covid19treatmentguidelines.nih.gov.

117 When Paxlovid was administered: Pfizer, "Pfizer's Novel COVID-19 Oral Antiviral Treatment Candidate Reduced Risk of Hospitalization or Death by 89% in Interim Analysis of Phase 2/3 EPIC-HR Study," Nov. 5, 2021, https://www.pfizer.com/.

120 According to the WHO: WHO, "COVID-19 Clinical Management/Living Guidance," Jan. 25, 2021, https://www.who.int.

120 Hundreds of thousands of people: Clinton Health Access Initiative, "Closing the Oxygen Gap," Feb. 2020, https://www.clintonhealthaccess.org/.

121 And as I write this: https://hewatele.org/.

122 Some 9,000 years ago: "Stone Age Man Used Dentist Drill," BBC News, April 6, 2006.

122 The ancient Egyptian physician: Rachel Hajar, "History of Medicine

Timeline," *Heart Views: The Official Journal of the Gulf Heart Association* 16, no. 1 (2015): 43–45.

123 Sometimes a drug has been invented: Alan Wayne Jones, "Early Drug Discovery and the Rise of Pharmaceutical Chemistry," *Drug Testing and Analysis* 3, no. 6 (June 2011): 337–44; Melissa Coleman and Jane Moon, "Antifebrine: A Happy Accident Gives Way to Serious Blues," *Anesthesiology* 134 (2021): 783.

126 In May 1747: Arun Bhatt, "Evolution of Clinical Research: A History Before and Beyond James Lind," *Perspectives in Clinical Research* 1, no. 1 (2010): 6–10.

129 It was ready to go: U.K. Research and Innovation, "The Recovery Trial," https://www.ukri.org.

133 Generic manufacturers produce: Center for Global Development, "Background Research and Landscaping Analysis on Global Health Commodity Procurement," May 2018, https://www.cgdev.org.

134 The WHO's malaria program: WHO, "Impact Assessment of WHO Prequalification and Systems Supporting Activities," June 2019, https://www.who.int.

134 Even in the United States: U.S. Food and Drug Administration, "Generic Drugs," https://www.fda.gov.

Chapter 6. Get ready to make vaccines

139 Figure: COVID vaccines were developed incredibly quickly. Year disease identified reflects when the respective virus was first isolated from patient samples. Vaccine availability marks first widely used vaccine for the respective disease. Global vaccination for whooping cough, polio, and measles represents share of one-year-olds who have been immunized against the disease. COVID-19 vaccinations include all eligible individuals through December 2021. Source: Samantha Vanderslott, Bernadeta Dadonaite, and Max Roser, "Vaccination" (2013), published online at OurWorldInData.org, retrieved from https://ourworldindata.org/vaccination. CC BY 4.0.

140 Historically, the average probability: Asher Mullard, "COVID-19 Vaccine Development Pipeline Gears Up," *The Lancet,* June 6, 2020.

141 In June, after seeing initial data: Siddhartha Mukherjee, "Can a Vaccine for Covid-19 Be Developed in Time?," *New York Times,* June 9, 2020.

141 The vaccine made by Pfizer: WHO, "WHO Issues Its First Emergency Use Validation for a COVID-19 Vaccine and Emphasizes Need for Equitable Global Access," Dec. 31, 2020, https://www.who.int.

141 To get a sense of just how fast: CDC, "Vaccine Safety: Overview, History, and How the Safety Process Works," Sept. 9, 2020, https://www.cdc.gov.

141 This remarkable feat: "Maurice Hilleman," Wikipedia, Dec. 2021.

142 Figure: Making a vaccine. Previously, the fastest a vaccine had been developed was four years (mumps), by Maurice Hilleman. The COVID time line of one year represents the time between the first effort to produce a COVID vaccine and the emergency authorization approval of Pfizer and BioNTech vaccine. Source: Reprinted with permission. N Engl J Med 2020; 382:1969–1973. Copyright 2020, Massachusetts Medical Society.

143 Photo (left): Paul Hennessy/SOPA Images/LightRocket via Getty Images; photo (right): Brian Ongoro/AFP via Getty Images.

144 Since 2000, it has helped: Gavi, "Our Impact," Sept. 21, 2020, https://www.gavi.org/.

145 Figure: Gavi saves lives. Cumulative number of children immunized with the last recommended dose of a Gavi-supported vaccine delivered through routine systems only, 2016–2020. Deaths under age five represent average probability of a child born in any of the Gavi-supported countries dying before reaching the age of five. Source: Gavi Annual Progress Report 2020; United Nations Inter-agency Group for Child Mortality Estimation 2021.

147 The number goes even higher: Joseph A. DiMasia et al., "Innovation in the Pharmaceutical Industry: New Estimates of R&D Costs," *Journal of Health Economics* (May 2016): 20–33.

148 By the summer of 2021: CEPI, "Board 24–25 June 2021 Meeting Summary," Aug. 19, 2021, https://www.cepi.net/.

149 COVAX was intended to solve: Benjamin Mueller and Rebecca Robbins, "Where a Vast Global Vaccination Program Went Wrong," *New York Times,* Oct. 7, 2021.

150 Illustration: The Gates Notes, LLC/Studio Muti.

153 In 1999, a cancer researcher: J. J. Wheeler et al., "Stabilized Plasmid-Lipid Particles: Construction and Characterization," *Gene Therapy* (Feb. 1999): 271–81.

153 Six years later, working: Nathan Vardi, "Covid's Forgotten Hero: The Untold Story of the Scientist Whose Breakthrough Made the Vaccines Possible," *Forbes,* Aug. 17, 2021.

154 Japan used *only:* "COVID-19 Vaccine Doses Administered by Manufacturer, Japan," Our World in Data, Jan. 2022, https://www.ourworldindata.org.

156 Table: Vaccines approved for WHO EUL as of January 2022. Data for estimated doses shipped are from Linksbridge Media Monitoring and UNICEF COVID-19 Vaccine Market Dashboard (accessed Jan. 2022), https://www.unicef.org.

158 The word *vaccine* comes from: Patrick K. Turley, "Vaccine: From *Vacca,* a Cow," U.S. National Library of Medicine, March 29, 2021, https://www.ncbi.nlm.nih.gov/.

158 That same year, a contaminated serum: "Antitoxin Contamination," *The History of Vaccines,* https://www.historyofvaccines.org/.

158 The job of regulation: "The Biologics Control Act," *The History of Vaccines,* https://www.historyofvaccines.org/.

159 The exploratory stage: "Vaccine Development, Testing, and Regulation," *The History of Vaccines,* Jan. 17, 2018, https://www.historyofvaccines.org/; "Phases of Clinical Trials," BrightFocus Foundation, https://www.brightfocus.org/.

165 In a second-source deal: Cormac O'Sullivan et al., "Why Tech Transfer May Be Critical to Beating COVID-19," McKinsey & Company, July 23, 2020, https://www.mckinsey.com.

167 In low-income countries, only 8 percent: Hannah Ritchie et al., "Coronavirus Pandemic (COVID-19)," Our World in Data, Jan. 2022, https://www.ourworldindata.org/.

167 Figure: Vaccine inequity. Population vaccinated represents the number of people who received at least one dose prescribed by the vaccination protocol. It does not include people having been infected with SARS-CoV-2. Source: Official data collated by Our World in Data. CC BY 4.0.

168 In 2021, the White House: "American Pandemic Preparedness: Transforming Our Capabilities," White House, Sept. 2021, https://www.whitehouse.gov/.

168 The most widely used one: "Indian Manufacturer Cuts Price of Childhood Vaccine by 30 Percent," Gavi, April 18, 2013, https://www.gavi.org/.

169 That's a sixteen-fold increase: Melissa Malhame et al., "Shaping Markets to Benefit Global Health—a 15-Year History and Lessons Learned from the Pentavalent Vaccine Market," *Vaccine: X,* Aug. 9, 2019.

169 And, as I was writing this book: "India Completes National Introduction of Pneumococcal Conjugate Vaccine," Gavi, Nov. 12, 2021, https://www.gavi.org/; "GBD Compare," IHME, https://www.healthdata.org/.

171 Figure: Global vaccination rates are higher than ever. WHO, Diphtheria tetanus toxoid and pertussis (DTP3), 2021 (accessed Jan. 2022); data provided by the World Bank Income Group: https://apps.who.int/gho/data. CC BY 4.0.

172 Photo: The Gates Notes, LLC/Uma Bista.

174 The measles vaccine: CDC, "Measles Vaccination," https://www.cdc.gov.

174 When Larry Brilliant: W. Ian Lipkin, Larry Brilliant, and Lisa Danzig, "Winning by a Nose in the Fight Against COVID-19," *The Hill,* Jan. 1, 2022.

177 Photo: The Gates Notes, LLC/Jason J. Mulikita.

Chapter 7. Practice, practice, practice

180 In July 2015, *The New Yorker:* Kathryn Schulz, "The Really Big One," *The New Yorker,* July 13, 2015.

181 The 2016 exercise involved: Washington Military Department, "Looking

at Successes of Cascadia Rising and Preparing for Our Next Big Exercise," June 7, 2018, https://m.mil.wa.gov; Emergency Management Division, "Washington State 2016 Cascadia Rising Exercise, After-Action Report," rev. Aug. 1, 2018, https://mil.wa.gov/.

182 As the WHO's flu preparedness program put it: WHO, "A Practical Guide for Developing and Conducting Simulation Exercises to Test and Validate Pandemic Influenza Preparedness Plans," 2018, https://www.who.int.

182 Credit for running: Karen Reddin et al., "Evaluating Simulations as Preparation for Health Crises Like CoVID-19: Insights on Incorporating Simulation Exercises for Effective Response," *International Journal of Disaster Risk Reduction* 59 (June 1, 2021): 102245.

183 Cygnus in particular highlighted: David Pegg, "What Was Exercise Cygnus and What Did It Find?," *The Guardian,* May 7, 2020.

183 The United States had a similar experience: U.S. Department of Health and Human Services, "Crimson Contagion 2019 Functional Exercise After-Action Report," Jan. 2020, accessed via https://www.governmentattic.org.

185 Less than two months later: Tara O'Toole, Mair Michael, and Thomas V. Inglesby, "Shining Light on 'Dark Winter,'" *Clinical Infectious Diseases* 34, no. 7 (April 1, 2002): 972–83.

189 In the summer of 2013: Kathy Scott, "Orland Int'l Battles Full-Scale Emergency (Exercise)," *Airport Improvement,* July–Aug. 2013.

189 At the other end: Sam LaGrone, "Large Scale Exercise 2021 Tests How Navy, Marines Could Fight a Future Global Battle," *USNI News,* Aug. 9, 2021.

190 A good model: Alexey Clara et al., "Testing Early Warning and Response Systems Through a Full-Scale Exercise in Vietnam," *BMC Public Health* 21, no. 409 (2021).

193 As my friend Nathan Myhrvold: Nathan Myhrvold, "Strategic Terrorism: A Call to Action," *Lawfare,* https://paper.ssrn.com.

196 In the early 1980s: Email correspondence with Bill Foege.

Chapter 8. Close the health gap between rich and poor countries

197 Across every age group: Samantha Artiga, Latoya Hill, and Sweta Haldar, "COVID-19 Cases and Deaths by Race/Ethnicity: Current Data and Changes over Time," https://www.kff.org.

197 In 2020, it pushed: Daniel Gerszon Mahler et al., "Updated Estimates of the Impact of COVID-19 on Global Poverty: Turning the Corner on the Pandemic in 2021?," *World Bank Blogs,* June 24, 2021, https://blogs.worldbank.org/.

198 In January 2021: Tedros Adhanom Ghebreyesus, "WHO Director-General's Opening Remarks at 148th Session of the Executive Board," Jan. 18, 2021, https://www.who.int.

198 "The Pandemic Has Split in Two": Weiyi Cai et al., "The Pandemic Has Split in Two," *New York Times,* May 15, 2021.

198 One WHO official denounced: James Morris, "Rich Countries Hoarding COVID Vaccines Is 'Grotesque Moral Outrage' That Leaves UK at Risk, WHO Warns," Yahoo News UK, May 6, 2021.

198 By the end of March: Our World in Data, "Share of the Population Fully Vaccinated Against COVID-19," https://www.ourworldindata.org.

199 Consider that by the end of 2021: Our World in Data, "Estimated Cumulative Excess Deaths During COVID, World," https://www.ourworldindata.org.

199 But compare it with: IHME, "GBD Compare," https://healthdata.org (accessed Dec. 31, 2021).

199 Figure: The health gap. Deaths per 100,000 population. High-income North America includes United States, Canada, and Greenland. Source: Institute for Health Metrics and Evaluation (IHME) at the University of Washington, Global Burden of Disease Study 2019.

200 A child born in the U.S.: "WHO, Life Expectancy at Birth (Years)," https://www.who.int.

202 Figure: More children are surviving today than at any other point in history. Under-five mortality data (5q0), the probability of dying between birth and exact age 5, is expressed as average annual deaths per 1,000 births. Source: United Nations, Department of Economic and Social Affairs, Population Division (2019), World Population Prospectus 2019, Special Aggregates, Online Edition, Rev. 1.

204 I admired the clever: Hans Rosling, "Will Saving Poor Children Lead to Overpopulation?," https://www.gapminder.org; Our World in Data, "Where in the World Are Children Dying?," https://ourworldindata.org/.

204 It happened in France: Bill and Melinda Gates Annual Letter, 2014, https://www.gatesfoundation.org/.

204 As the United Nations Population Fund explains: "Demographic Dividend," https://www.unfpa.org/.

208 It raised almost $4 billion: The Global Fund, "Our COVID-19 Response," https://www.theglobalfund.org (accessed Dec. 2021).

208 Even though about a sixth: WHO, "Tuberculosis Deaths Rise for the First Time in More Than a Decade Due to the COVID-19 Pandemic," Oct. 14, 2021, https://www.who.int.

209 Part of its mission: Gavi, https://www.gavi.org.

210 After it won independence: Chandrakant Lahariya, "A Brief History of Vaccines & Vaccination in India," *Indian Journal of Medical Research* 139, no. 4 (2014): 491–511.

210 In 2000, for example: WHO Immunization Dashboard for India, https://immunizationdata.who.int/.

211 Figure: Stamping out measles in India. Measles vaccines include first dose (MCV1) and second dose (MCV2). Annual number of measles cases include clinically confirmed, epidemiologically linked, or by laboratory investigation. Source: WHO, Measles vaccination coverage, 2021 (accessed Jan. 2022), data reported through the WHO/UNICEF Joint Reporting Form on Immunization and the WHO/UNICEF Joint Estimates of National Immunization Coverage: https://immunizationdata.who.int. CC BY 4.0.

212 Within a few weeks: Global Polio Eradication Initiative, "The First Call," March 13, 2020, https://polioeradication.org/.

212 Staff from the polio EOC: Interview with Faisal Sultan, Oct. 13, 2021.

212 By late summer 2021: Our World in Data, "Daily COVID-19 Vaccine Doses Administered per 100 People," https://ourworldindata.org/.

214 In 2019, the answer: IHME, "Flows of Development Assistance for Health," https://vizhub.healthdata.org.

214 People spend almost that much: Statista Research Department, "Size of the Global Fragrance Market from 2013 to 2025 (in Billion U.S. Dollars)," Nov. 30, 2020, https://www.statista.com.

214 Figure: Child deaths cut in half. Total deaths from communicable, neonatal, and nutritional diseases for children under age 5 years old, 1990–2019. Source: Institute for Health Metrics and Evaluation (IHME) at the University of Washington, Global Burden of Disease Study 2019.

215 Figure: Going after preventable diseases. Deaths under age 5 from select preventable causes. Deaths from pneumonia represent "lower-respiratory infections." Source: Institute for Health Metrics and Evaluation (IHME) at the University of Washington.

Chapter 9. Make—and fund—a plan for preventing pandemics

223 The last known smallpox death: CDC, "History of Smallpox," https://www.cdc.gov.

229 In many low-income countries: The Primary Health Care Performance Initiative, https://improvingphc.org/.

232 Because the world has: G20 High Level Independent Panel on Financing the Global Commons for Pandemic Preparedness and Response, "A Global Deal for Our Pandemic Age," June 2021, https://pandemic-financing.org.

232 The idea that rich countries: OECD, "The 0.7% ODA/GNI Target—a History," https://www.oecd.org.

Afterword: How COVID changed the course of our digital future

237 In the U.S., for example: Pew Research Center, "Mobile Fact Sheet," https://www.pewresearch.org.

238 And just about everyone started: U.S. Census Bureau, "Quarterly Retail E-Commerce Sales, 4th Quarter 2020," Feb. 2021, https://www.census.gov.

244 The number of people using: Oleg Bestsennyy et al., "Telehealth: A Quarter-Trillion-Dollar Post-COVID-19 Reality?," McKinsey & Company, July 9, 2021, https://www.mckinsey.com/.

247 Algebra I is a key milestone: Timothy Stoelinga and James Lynn, "Algebra and the Underprepared Learner," UIC Research on Urban Education Policy Initiative, June 2013, https://mcmi.uic.edu/.

248 The gap has narrowed: Emily A. Vogels, "Some Digital Divides Persist Between Rural, Urban and Suburban America," Pew Research Center, Aug. 19, 2021, https://www.pewresearch.org/.

248 This is especially true for: Sara Atske and Andrew Perrin, "Home Broadband Adoption, Computer Ownership Vary by Race, Ethnicity in the U.S.," Pew Research Center, July 16, 2021, https://www.pewresearch.org.

249 Photo: AT&T Photo Service/United States Information Agency/PhotoQuest via Getty Images.

INDEX

Page numbers in *italics* refer to illustrations and captions.

A NOTE ON THE TYPE

This book was set in Adobe Garamond. Designed for the Adobe Corporation by Robert Slimbach, the fonts are based on types first cut by Claude Garamond (ca. 1480–1561).

Composed by North Market Street Graphics,
Lancaster, Pennsylvania

Printed and bound by Berryville Graphics,
Berryville, Virginia

Designed by Michael Collica